FRONTIERS OF ILLUSION

FRONTIERS OF

ILLUSION

Science, Technology,

and the Politics

of Progress

Daniel Sarewitz

TEMPLE UNIVERSITY PRESS

PHILADELPHIA

Temple University Press, Philadelphia 19122
Copyright © 1996 by Daniel Sarewitz. All rights reserved
Published 1996
Printed in the United States of America
♾ The paper used in this book meets the requirements of
the American National Standard for Information Sciences –
Permanence of Paper for Printed Library Materials,
ANSI Z39.48–1984.
Text design by Richard Eckersley
Library of Congress Cataloging-in-Publication Data
Sarewitz, Daniel R.
Frontiers of illusion : science, technology, and the politics of
progress / Daniel Sarewitz.
p.   cm.
Includes bibliographical references and index.
ISBN 1-56639-415-5 (cloth : alk. paper). – ISBN 1-56639-416-3
(pbk. : alk. paper)
1. Science and state – United States. 2. Technology and state –
United States. 3. Research – Philosophy. 4. Research –
Government policy – United States. 5. Research –
Social aspects – United States. I. Title.
Q127.U6S25 1996   338.97306 – dc20   95-31840

For Erica

CONTENTS

CONTENTS

PREFACE

THE CONCERNS OF THIS BOOK must be clearly distinguished from those of a persistent and sometimes vitriolic academic debate that revolves around the character and validity of the natural sciences. The debate hinges on the core assertion of science: that some things are objectively knowable. Is scientific knowledge a reflection of the true structure of an objective outside world, gradually and progressively approached through the scientific method? Or, rather, does the inescapable cultural context in which experiments and observations are carried out permit nothing more than the construction of a vague and distorted simulacrum of nature, a reflection more of ourselves than of the world that surrounds us? The two sides of the debate are variously characterized as science versus antiscience, modernist versus postmodernist, natural science versus social science, objectivist versus relativist.

Many books and articles have emerged from the dispute, but this book is not one of them. Nevertheless, because the arguments that will be presented here are critical of certain political and social implications of science and technology in modern culture, those who take issue with my critique may be inclined to interpret it within the context of the philosophical controversy about the intrinsic validity of science. To do so is to miss utterly the central concern of this book, which has nothing to do with what science is and everything to do with what science does and can do once it emerges from the laboratory. Therefore, to avoid confusion and misrepresentation, I here baldly and unapologetically state that I recognize the scientific method to be a valid technique for approaching what I am pleased to term an objective understanding of the physical and natural world. This belief, however, offers no a priori comfort to anyone who would try to answer such questions as What types of scientific knowledge should society choose to pursue? How should such choices be made and by whom? How should society apply this knowledge, once gained? How can "progress" in science and technology be defined and measured in the context of broader social and political goals? And indeed, it is precisely these sorts of questions that underlie and motivate this book.

ix

Yet to ask such questions is to provoke accusations of antiscientism. Commenting on an early draft of the book, an anonymous physicist wrote: "[The] important policy question is not whether social progress follows inevitably from scientific research . . . but rather whether progress is possible without research. From that starting point one is led to explore paths that Sarewitz hasn't begun to imagine." When critics reduce the "important policy question" to two antithetical alternatives—research or no research; science or antiscience—the overall theme here under consideration, how research can best serve society, is transmogrified into a choice between the status quo and nothing at all. This, of course, is a fine choice for the beneficiaries of the status quo.

No reasonable person can fail to appreciate that science and technology play a crucial role in modern society, and that future prospects for societal well-being on a national and global scale will depend in part upon further scientific research. This appreciation does not, however, demand conformity or complacency in assessing the relationship between scientific progress and the welfare of humanity. Every important modern cultural institution is evolving under the influence of science and technology. The very nature of democratic society demands, therefore, that this influence be subject to a rigorous process of dialogue and critique. Debate about the prevailing attitudes and institutions that govern research and development is no more antiscientific than political debate is antidemocratic. The question is not "do we need science?" but "what science do we need?"

Although the origins of this book cannot be traced to any particular moment in my past, it would never have been written were it not for the Congressional Science and Engineering Fellowship program of the American Association for the Advancement of Science, which brought me (and twenty-some other morbidly curious scientists) to Washington, D.C., on a one-year fellowship in 1989, and thus allowed me to make the transition from academic basic research in the earth sciences to science and technology policy in the U.S. Congress. In July I was studying the connection between deformation of the earth's crust and earthquakes in the Peter the First Range of Tadjikistan; three months later I was writing earthquake-preparedness legislation in the wake of the temblor that had struck the San Francisco Bay region. I was hooked; against all my better instincts, I stayed in Washington for almost four years, first as a fellow in the office of Congressman George E. Brown, Jr., and then as science consultant to Congressman Brown when he became chairman of the House Committee on Science, Space, and Technology. This book is one product of those years, an effort to integrate two worldviews—the political and scientific—that may seem antithetical yet must somehow be reconciled if the goals of either are to be achieved.

The complete manuscript was reviewed by Erica Rosenberg, Radford Byerly, Jr., and Yosef Grodzinsky; individual and multiple chapters were reviewed by Daryl Chubin, Ronald Brunner, Richard Barke, Charles Blahaus, Virgil Frizzell, Roger Pielke, Elizabeth Knoll, and four anonymous readers. Their comments, criticisms, challenges, and support were crucial to the book's evolution; I am deeply grateful for their help. Special thanks go to George Brown and Rad Byerly for creating a wonderfully stimulating and challenging intellectual environment in that most unlikely of all settings: the United States House of Representatives.

FRONTIERS OF ILLUSION

1

The End of the Age of Physics

When human life lay grovelling in all men's sight, crushed to the earth under the dead weight of superstition whose grim features loured menacingly upon mortals from the four quarters of the sky, a man of Greece was first to raise mortal eyes in defiance, first to stand erect and brave the challenge. Fables of the gods did not crush him, nor the lightning flash and the growling menace of the sky. Rather, they quickened his manhood, so that he, first of all men, longed to smash the constraining locks of nature's doors. The vital vigour of his mind prevailed. He ventured far out beyond the flaming ramparts of the world and voyaged in mind throughout infinity. Returning victorious, he proclaimed to us what can be and what cannot: how a limit is fixed to the power of everything and an immovable frontier post. Therefore superstition in its turn lies crushed beneath his feet, and we by his triumph are lifted level with the skies.—Lucretius, *On the Nature of the Universe*

THE AGE OF PHYSICS came to an end on October 21, 1993, when the U.S. Congress canceled funding for the Superconducting Super Collider, a $10 billion-plus project whose scientific goals were to reproduce the conditions of the very earliest stages of the big bang and determine the origin of mass in the universe. To its supporters, the collider was many things: it was the ultimate physics experiment, an opportunity to develop a "final theory" that could unify all known laws of physics; it was a testament to humanity's quest for knowledge and enlightenment and a laboratory for the education of America's next generation of physicists; it was a further demonstration of American preeminence in science and technology as well as a potential source of technological innovation that could yield

1

great economic and practical value; and it was a sizeable public works project that would create thousands of jobs and bring billions of dollars to the Texas economy. But despite these justifications, despite the political support that they engendered, and despite the $2 billion that the federal government had already spent on construction, the collider was canceled. Today, fourteen miles of the planned fifty-four-mile circular tunnel beneath the ground of Waxahachie, Texas, is all that remains of the project, a subterranean monument to the fifty-year period where physics reigned supreme in American science and technology policy.

Many explanations have been offered for the demise of the Superconducting Super Collider (SSC), all of which probably contain an element of truth. Highly publicized cost overruns and poor project management at the Department of Energy created political vulnerability just at a time when a beleaguered Congress was looking to make symbolic, if not significant, cuts in federal spending. This vulnerability was reinforced by the political impotence of the Texas congressional delegation and by a conspicuous lack of consensus about the value of the SSC within the scientific community itself. Foreign governments, whose support and cooperation were supposedly a cornerstone of the project, proved willing to contribute little more than rhetoric on the SSC's behalf. Ultimately, and most importantly, the proponents of the SSC failed to convince Congress and the public that the project was crucial to the future of science and the nation.

And yet, five years earlier Congress would never have killed a major physics research program with as high a profile as the SSC's. Physics was the discipline that helped win World War II, and physics helped to keep the Soviet Union at bay throughout the Cold War. Physicists were the most influential scientists in the nation, garnering Nobel Prizes, advising presidents, and making the discoveries that allowed the United States to lead the world in technological innovation. Physicists even acted as the conscience of the nation by spearheading the opposition to a nuclear arms race that they had unwittingly unleashed. Embodied by the benevolent image of Albert Einstein, physicists represented the intelligence, ingenuity, and wisdom that brought peace and prosperity to America in the twentieth century, and the SSC represented the type of fundamental investigation into nature that physicists had long promoted as the key to progress in modern society.

By the early 1990s, however, several historical trends had undermined the special political treatment long enjoyed by physicists. Perhaps most important was the end of the Cold War and the consequent disappearance of the principal political rationale for support of physics research in America—national defense. Another factor was the protracted anemia of the American economy and the apparent inability of the nation to translate scientific and technological progress into economic growth. Also implicated was the rising political appeal of other scientific disciplines, especially biomedical research, which seemed to offer more to society than physics did. Each of these trends was a harbinger of a new and evolving government research agenda. While the SSC lost its mandate, other major science programs were thriving, including the Human Genome Project; special government initiatives in manufacturing technology, high performance computing, and biotechnology; and the federal research program on global climate change.

Despite such changes, the policies, programs, and attitudes that were institutionalized during the age of physics still exert a profound influence on the nation. The mushroom clouds over Japan cast a long shadow, and while this shadow darkened the prospects of world peace, it also symbolized the imagination, potential, and potency of all modern science. The promise of science and technology, borne out so tangibly and spectacularly during World War II, translated into a tremendously expanded role for government in promoting science and technology in the United States after the war. Federal funding for research and development grew from less than $2 billion per year in the early 1950s to more than $70 billion per year in the early 1990s.[1]

Yet the past fifty years have shown that the linkage between scientific progress and societal well-being is highly attenuated. While federally funded research and development (R&D) has historically been justified by its promised contribution to human welfare, the fulfillment of this promise seems increasingly elusive. Knowledge and innovation grow at breathtaking rates, and so does the scale of the problems that face humanity. Science-based revolutions in areas such as communication and information technologies, agriculture, materials, medical technology, and biotechnology are accompanied by global weapons proliferation, population growth, concentration of wealth, declining biodiversity and loss of habitat, deterioration of arable land, destruction of stratospheric ozone, and the potential for rapid changes in the earth's climate. These opposing trends may appear to have little possible connection to one another, but they create at the very least an arresting counterpoint, a fundamental paradox of modern culture.

Nevertheless, advocates of publicly funded science—not only researchers themselves but also politicians, bureaucrats, university administrators, corporate executives, special-interest groups, and many private citizens—base their advocacy on the premise and promise that more scientific knowledge and technological innovation will lead to the solution of society's most serious challenges. This line of argument uniformly assumes a causal linkage between progress in science and technology and progress in society. As Nobel Prize–winning physicist Leon Lederman suggests: "What's good for American science . . . is good for America."[2]

The perspective adopted throughout this book is that government support for R&D must ultimately be justified by the creation of societal benefits. Modern society is obviously dependent in many ways on science and technology, and the federal government has helped to create the world's most advanced system of research and development in response to this dependence. But the R&D system is therefore a political entity, itself dependent upon government decision-making processes and public approval for its own well-being. In this context of dependence, and in light of the growing complexity and magnitude of challenges to humanity's long-term welfare, the assertion of causality between progress in the laboratory and progress in society may therefore be viewed as an unproven—although extremely powerful—political argument invoked by researchers and research advocates to sustain public support. Upon such arguments the research system is built.

This book is a critique of the foundations of United States science and technology policy at the end of the age of physics and a portrayal of the R&D system in its political setting. It is not a critique of science and technology per se,* nor is it a discussion of the byzantine and frequently capricious political process through which policy is implemented. Rather,

---

*The word "science" as used in this book refers to a body of existing and prospective knowledge, not to the technical activity of scientific research or the epistemological founda-

the goal here is to reveal and evaluate the assumptions and interests that underlie the current system and that justify the government's role in supporting and promoting research and development. These assumptions and interests help to shape the complex relationship between scientific progress and the common good.

Science and Technology Policy

In 1995, the U.S. government spent almost $73 billion on science and technology programs. How much money is this? It is considerably less than was spent on the military ($276 billion) and on interest on the national debt ($234 billion) and considerably more than was spent on education and training programs ($56 billion), administration of the judicial system ($18 billion), foreign aid ($13 billion), and pollution control and abatement ($6 billion). Fifty-four percent of the science and technology budget was devoted to military projects; the rest was distributed throughout the gamut of civilian research and development fields ranging from space exploration and energy supply to biomedical studies and agriculture.

The total realm of government-funded science and technology activities is loosely encompassed by the term "federal research and development system." The processes through which this system is designed, promoted, funded, administered, and evaluated are encompassed—equally loosely—by the term "science and technology policy." Together, these terms offer a convenient and unavoidable rubric for the full range of activities that determine the shape, size, and direction of publicly funded science and technology in America. The "system" is in fact pluralistic; balkanized might be a more precise description. It is made up of many competing elements, including congressional committees and federal

---

tions of science. This perspective is external to the laboratory and reflects a social consensus that treats the validity of the scientific method as proven.

As a matter of taxonomy, the word "science" is meant to encompass all of the natural sciences—for example, physics, chemistry, biology, astronomy, earth sciences—but to exclude the social sciences. References to or discussions of the social sciences will be explicitly identified. Engineering is subsumed by the term "technology," although engineering activities are a component of both "research" and "development."

agencies fighting amongst themselves for greater influence over budgets and programs; national weapons laboratories desperately trying to justify their continued existence in the post–Cold War world; universities seeking to establish or maintain world-class facilities; and individual scientists conducting research and competing for research funds. The "policy process," which determines the structure of the system, is perhaps best characterized as the total of the tactics and strategies employed by these various competing elements as they strive to further their own interests. This process sounds unsystematic and Darwinian, and in many ways it is.

Traditional science and technology policy covers the R&D trinity: basic research, applied research, and development. Basic research is the investigation of natural phenomena; it is often referred to as "pure science" because it is supposedly carried out independently from any consideration of practical utility. The popular (and often apocryphal) imagery of science—Archimedes shouting "Eureka!" in his bathtub, Galileo dropping things off the Leaning Tower of Pisa, Fleming accidentally discovering the penicillin mold—is the imagery of basic research. Applied research is the effort to use existing knowledge to solve particular problems— for example, determining the medical utility of penicillin. Development is concerned with making the products of applied research usable by society—for example, by devising techniques for mass production of penicillin. The boundaries between these various activities are often diffuse, but as a whole they encompass the jurisdiction of science and technology policy. In recent decades, between 50 and 60 percent of the federal R&D budget has been spent on technology development, predominantly for military programs administered by the Pentagon. Most research programs, in contrast, are focused on nonmilitary goals and are administered by a variety of federal agencies, including the National Institutes of Health, National Science Foundation, National Aeronautics and Space Administration, and the Department of Energy.

Because the federal government supports almost half of the nation's R&D activities (private corporations support most of the remainder), the tortuous policy process by which government research priorities and spending levels are determined is the single most important factor controlling the overall makeup of the nation's R&D system. How much will be spent on energy research and how much on medical research? Within the energy research budget, how much is allocated to solar power and how much to nuclear power? How much should go for applied research on nuclear reactors and how much for basic research in high-energy physics? These types of questions are the ultimate responsibility of politicians and bureaucrats. As with all areas of government expenditure, however, various interest groups try to influence the way that R&D programs are funded and administered. The principal interest group in this case is the research community itself—especially the scientists and engineers who carry out federal R&D programs—although, once again, the term "community" in fact comprises a diverse group with diverse and often competing interests. Still, researchers can be distinguished from other participants in the R&D policy process by their technical expertise. They know how to set up laboratories and research programs, they understand the potentials and limits of their work, and they possess comprehensive knowledge in their particular fields of endeavor—knowledge that is often far too technical in nature for nonscientists to grasp fully. Many of the same scientists and engineers who coordinated technical activities for the government during World War II were later called upon to design the postwar R&D system. In fact, for the past fifty years, the government has depended upon the advice of researchers in formulating and implementing federal science and technology policy.

As a political matter, the government spends over $70 billion a year on R&D not because it cares about the abstract pursuit of knowledge but because policy makers and the public believe that progress in science and technology creates societal benefits, both tangible and intangible. Military preparedness, economic growth, medical care, adequate energy resources, and national prestige are among the most important benefits said to accrue from the new knowledge and innovation that come from the R&D system.* Because the principal political justification for R&D funding is public benefit, the basis for science and technology policy can be thought of as a social contract under which the government provides

*The value of science as a cultural activity—a source of intellectual liberation, spiritual enlightenment, and aesthetic satisfaction—is incalculable, but it is not a principal motivation for government support of research and development and is therefore generally beyond the scope of this book.

support for R&D activities in return for a product—knowledge and innovation—that contributes to the common good.

Policy Goals and Policy Myths

But how is public benefit created from the laws of nature? The remarkable success of physics derives principally from its capacity to isolate particular phenomena and describe them in terms—usually mathematical—that are applicable to a large number of situations. The archetypal example is Newton's laws of motion, three basic principles that are valid at scales varying from interstellar to molecular and are applicable to such apparently disparate phenomena as heat and temperature, electric and magnetic forces, celestial mechanics, and motion due to the earth's gravity. Similarly, four simple equations derived by James Clerk Maxwell in the 1860s successfully unified and described all electromagnetic phenomena and thus set the stage for the electronics revolution a century later. The beauty and strength of these sorts of simplifying relationships is that they create a view of the world that is uniform and invariant—a world that is independent of context and therefore describable and predictable.

By reducing natural phenomena to a series of ideal, universal relationships and controlling the environment in which these phenomena occur, scientists and engineers are able to extract practical utility from the laws of nature. It is through such a reductionist approach to creating and applying knowledge that the natural sciences—physics, biology, chemistry, and their various offspring—have made their greatest impact on society. Thus, for example, a physicist can derive an equation to describe the behavior of an electron moving through a uniform electric field in a vacuum. Such ideal conditions are not easily found in nature, but a physicist can create them in the laboratory. The result is a cathode ray tube, and one of the linear descendents of the cathode ray tube is television. Similarly, physicists know, through Einstein's theory of special relativity, that the destruction of atomic mass liberates huge quantities of energy. Fortunately, this conversion occurs naturally on earth only in very small doses. However, by concentrating relatively small amounts of highly unstable elements such as uranium 235 and bombarding the uranium with neutrons, physicists are able to achieve very large and self-sustaining mass-to-energy conversions indeed, and the result is nuclear weapons.

When a new process or product emerges from the laboratory, it undergoes a profound transition—from well-behaved, insular idea or object to dynamic component of a complex interactive social system. Once imbedded in that social system, the new idea or innovation may produce effects that are completely surprising. When a television is turned on, a series of intrinsically predictable electromagnetic processes occurs inside the television that always leads to the generation of a visual image on the screen. But nothing else about the television is predictable or immanent because all of its other attributes derive not from the physical laws that allow it to operate but from the context in which it is used: when, where, and by whom it is turned on; what is being broadcast; how the viewer is affected by the program; what activities the viewer chose to forego in making the decision to watch; how this decision affected others who interact with the viewer (a sports-hating spouse on Superbowl Sunday, for example); how the total number of viewers influences the economic prospects of companies that are advertising at that particular time.

On one hand there are laboratories and on the other, the outside world of human beings. An idea or product in the laboratory often evolves rapidly into something entirely different once it moves into society. But society, too, will undergo change as it responds to and absorbs the knowledge and innovation transmitted from the laboratory. No one anticipated the overwhelming impact of television—just a modified cathode ray tube—on global culture. Or of the telephone, which in its early days was viewed as little more than a replacement for the telegraph, useful only for point-to-point communication, like two cans and a piece of string. Computers were considered by their earliest designers to have application for a narrow range of scientific calculations and no commercial market appeal whatsoever.[3] Nuclear power, on the other hand, was to be the miracle that would provide electricity to the world at virtually no cost.[4] The societal value of science and technology is created at the interface between the laboratory and society; it is inherent in neither alone. There is no a priori link between the structure of physical reality and the manner in which society applies its scientific understanding of that structure.

And this is where politics collides with the natural world to define the fundamental dilemma of science and technology policy. Effective political justifications for government support of R&D are based primarily upon the promise of social benefit. But the societal consequences of scientific and technological progress are not inherent in the natural laws that researchers seek to uncover and exploit. The laws of nature do not ordain public good (or its opposite), which can only be created when knowledge and innovation from the laboratory interact with the cultural, economic, and political institutions of society. Modern science and technology policy is therefore founded upon a leap of faith: that the transition from the controlled, idealized, context-independent world of the laboratory to the intricate, context-saturated world of society will create social benefit. And society has obviously been willing to take the leap, as the annual $70 billion in public expenditures on research and development—or any visit to a hospital, electronics store, or guided missile cruiser—will attest.

An examination of the political rhetoric used to justify and explain the structure of the R&D system yields a number of powerful and oft-heard arguments that underlie and rationalize the leap of faith. Much of this book will be devoted to a discussion of these arguments. They are the basis for postwar science policy, and as such they have proven sufficiently resilient and compelling to ensure the political acceptability of the publicly funded R&D system. These arguments are characterized here as "myths" because they are widely subscribed to and commonly repeated, even though they are not derived from well-developed empirical or theoretical foundations. They are, at root, expressions of ideology and tools of political advocacy, accepted and expressed as truth. They guide the behavior of scientists and policy makers alike. Five such myths are identified and discussed in the following chapters:

1. The myth of infinite benefit: More science and more technology will lead to more public good.
2. The myth of unfettered research: Any scientifically reasonable line of research into fundamental natural processes is as likely to yield societal benefit as any other.
3. The myth of accountability: Peer review, reproducibility of results, and other controls on the quality of scientific research embody the principal ethical responsibilities of the research system.

4. The myth of authoritativeness: Scientific information provides an objective basis for resolving political disputes.
5. The myth of the endless frontier: New knowledge generated at the frontiers of science is autonomous from its moral and practical consequences in society.

These myths reflect the frame of reference of the laboratory. They are the starting point for political debate and thus are not subjected to scrutiny or analysis. Broad acceptance of the myths in the political realm can be attributed to three factors. First, the myths are wielded by a research community that possesses significant political legitimacy and enjoys high societal prestige, in large part because of the great and conspicuous progress of science and technology during this century. Second, the experience of rising standards of living in industrial society over the past two centuries is at least partly consistent with the assertion of causality between public welfare and progress in science and technology. Third, the political interests of the R&D community may overlap with the political interests of other groups, such as the manufacturing sector, the military, and even elected officials, who are thus willing to support or even co-opt the myths. Each of these considerations will be recurrent themes in coming chapters.

Because the overriding political rationale for government support of R&D has been the creation of public benefit, the myths of science and technology policy can be evaluated in this same context. But how are the criteria for such an evaluation established? Indeed, how can such general terms as "public benefit" be usefully defined? Myths of course serve a vital function in any society, by simplifying complex processes and making them comprehensible, and by encapsulating broadly held sentiments and thus creating a unity of vision and a shared sense of community and purpose. Thus, the question here is not simply the degree to which the policy myths are "true" or "false" but what ends they serve and how they affect the policy process and society at large. Just as the R&D community both defines and portrays itself and its interests through its particular myths, so are the broadest interests of society given definition, consistency, and resilience through shared civic myths of a higher order: "that all men are created equal; that they are endowed by their Creator with certain inalienable rights; that among these are life, liberty, and the pur-

11

suit of happiness." Such myths may be idealized and unachievable, and their interpretation may change with time, yet they embody particular goals toward which progress can be measured: to "establish justice, insure domestic Tranquility, provide for the common defense, promote the general Welfare, and secure the Blessings of Liberty to ourselves and our Posterity." Furthermore, the myths may define concrete standards from which society should not deviate: freedom of speech, of religion, of the press, of public assembly. Thus, the higher-order civic myths, in defining the fundamental, shared aspirations of society, create a framework within which the myths of science and technology policy can be understood and assessed.

The political effectiveness of the policy myths does not require that all scientists and engineers subscribe to them uncritically. However, the leading voices of the R&D community explicitly proffer these myths in their efforts to explain and justify the operations of the R&D system. In this sense, the public articulation of the myths represents the way that this community wishes to be viewed by the rest of the world, and it presumably represents, in a general manner, the way that the community views itself. For this reason, the presentation and analysis of policy myths in this book is based on the words chosen by leaders of the R&D community to portray themselves to the outside world and to each other. These words are found in newspapers and popular magazines, on the editorial and letters pages of scientific journals, in speeches and congressional hearings, and in the reports and publications of professional societies and other organizations that represent the interests and expertise of scientists and engineers. Such sources are explicitly nontechnical in nature; they are the public voice of science and technology. This is the voice heard by the rest of society.

Although much scholarly work in the social sciences over the past several decades has been devoted to analysis of the science-technology-society relationship, most of that work is of an academic nature that is not directly applicable to political debate; nor does it affect public perceptions about R&D. The scholarly literature is for the most part, therefore, separate from my principal concern here, which is to address the terms of political debate and public perception head-on. In fact, some of the myths still adhered to by natural scientists and policy makers alike have been assessed (and often debunked) in the social sciences literature;[5] what matters here, however, is that the myths remain powerful in the policy arena, that they are still part of the policy rhetoric, and that they have a palpable influence on policy decisions and thus on human welfare.

Beyond the Age of Physics: Science, Technology, and Reality

It is perhaps no coincidence that the end of the age of physics comes at a time when the human perspective has become truly global. Is the reductionist approach to understanding reality a suitable basis for confronting the problems of an increasingly interconnected world? What start out as concrete scientific and technological solutions to particular challenges—food production, birth control, energy supply, infectious diseases, transportation—are later revealed as interlocking components of a global society. Improvements in human health lead to exploding population; stronger and more durable materials and tools allow more efficient depletion of natural resources; greater industrial productivity generates more waste; techniques for increased food production introduce new toxic materials into the soil and water and reduce genetic variability of seed stocks. Further advances in science and technology will help to mitigate these problems, and they will also create new problems of their own. Context that is stripped away in the laboratory reasserts itself with a vengeance when a new process or innovation interacts with political, economic, and cultural systems. At the science-society interface, the distinction between solutions and problems may become fuzzy, as short-term scientific and technological contributions to human welfare often create unanticipated long-term problems. This trend is well known.

The history of human culture can be viewed as a history of crises, overcome in many cases with the help of technology—crises of demography, of resource supply, of environment, of economics, of military confrontation, of epidemic disease—but all such crises, with the exception of the Pleistocene ice age, were limited in their consequences by the finite scope and reach of society. Greece fell, but Rome rose; Europe maundered in the Dark Ages while Aztec, Mayan, Islamic, and Chinese cultures variously thrived. This long-term equilibrium may be a source of modern optimism—we have seen, and surmounted, serious challenges before.

Prophets of doom have come and gone; humanity's reach still expands. But there is something new on the scene: a quantum leap in the scale of human activities and human problems, a leap made possible by science and technology, with implications that can no longer be contained by political or geographic boundaries, or even by the natural systems that sustain life.

As the potency of science and technology grows, and the global interdependence of society deepens, the need to view scientific and technological progress in its human context becomes ever more urgent. Humanity's ever-increasing capacity for manipulation of nature at a fundamental level implies an accelerating potential for complex and even chaotic results at a societal level. No major product of the R&D system finds its way into society without in some way influencing or altering the economic, political, environmental, or even moral composition of life. Globalization of communication, of markets, of conflict, of environmental impacts, of culture, has magnified the nonlinear consequences of the research and development process and increased the opportunity both for major societal gain and for large-scale disaster.

The survival value of a linear policy perspective—where progress in science is viewed simply as the precursor of progress in society—will continue to diminish as the potency of science and technology increases. An essential responsibility of science and technology policy, therefore, must be to compare the promises made on behalf of the R&D system—promises that derive their legitimacy from the policy myths—with the system's actual performance in society, and to modify the promises and the myths on one hand, and the system on the other, to create greater consistency, a more realistic level of expectations, and an increased capacity to achieve societal goals. A new perspective must emerge in which science and technology are understood to be agents of context alteration, not merely simple steps upon which humanity seeks to ascend above the latest array of crises. Is the impact of the internal combustion engine on society best understood in terms of the thermodynamics of combustion, the average number of vehicle miles traveled each year, the contribution of the automotive industry to the national economy, the curve of rising carbon dioxide content in the atmosphere, the geopolitical struggle for control of oil resources embodied by the 1991 Persian Gulf War, or the effect of the commuter culture and the increased mobility of the individual on the structure and psyche of the American community? Understanding the implications of progress may become as important to the future of humanity as progress itself.

What follows is an attempt to illuminate and assess the foundations of post–World War II science and technology policy through an examination of the fundamental myths that have guided policy making for half a century (Chapters 2–6) and a discussion of some of the less familiar and more troubling economic and political ramifications of the research and development system that is constructed upon these myths (Chapters 7 and 8). This analysis will point the way toward some alternative approaches to science and technology policy, aimed at encouraging the development of stronger linkages between the publicly funded R&D agenda and the long-term goals and obligations of society (Chapter 9). The leap of faith that spans the chasm between laboratory and reality must be replaced with a bridge, lest, at the end of the age of physics, we look down and realize that there is nothing underneath our feet.

2

The Myth of Infinite Benefit

There is a concept which corrupts and upsets all others. I refer not to Evil, whose limited realm is that of ethics; I refer to the infinite.—Jorge Luis Borges, "Avatars of the Tortoise," *Labyrinths*

THE LANGUAGE USED to portray the expected benefits of scientific research has not changed much since 1945, when Vannevar Bush, director of the wartime Office of Scientific Research and Development under presidents Roosevelt and Truman, issued his famous report, *Science, the Endless Frontier*.[1] This document virtually codified the rationale for government support of R&D in the post–World War II era, and in doing so created a rhetorical template for explaining the value of science and technology in modern society:

> Progress in the war against disease depends upon a flow of new scientific knowledge. New products, new industries, and more jobs require continuous additions to knowledge of the laws of nature, and the application of that knowledge to practical purposes. Similarly, our defense against aggression demands new knowledge so that we can develop new and improved weapons. This essential, new knowledge can be obtained only through basic scientific research. . . . Advances in science when put to practical use mean more jobs, higher wages, shorter hours, more abundant crops, more leisure for recreation, for study, for learning how to live without the deadening drudgery which has been the burden of the common man for ages past. But to achieve these objectives . . . the flow of new scientific knowledge must be both continuous and substantial.[2]

Nearly half a century later, the federal government was using updated versions of the same language:

> Science does indeed provide an endless frontier. . . . The unfolding secrets of nature provide new knowledge to address crucial challenges, often in unpredictable ways. These include improving human health, creating breakthrough technologies that lead to new industries and high quality jobs, enhancing productivity with information technologies and improved understanding of human interactions, meeting our national security needs, protecting and restoring the global environment, and feeding and providing energy for a growing population.[3]

Such arguments—familiar to the point of banality—are integral to the political and economic philosophy of the modern state. And they carry with them an explicit corollary: if science and technology are important to the well-being of society, then the more science and technology society has, the better off it will be. Thus, a 1947 report by John Steelman, chairman of President Truman's Scientific Research Board, urged that "as a Nation, we increase our annual expenditures for research and development as rapidly as we can expand facilities and increase trained manpower."[4] Indeed, from 1960 to 1990 total national expenditures (government and industry) on R&D grew by 250 percent, while federal funding for civilian research and for university research (civilian and defense-related) grew by more than 400 percent (*after* inflation).[5] Calls for more growth continued. In 1992, President Bush advocated a doubling of the budget of the National Science Foundation (NSF) over a five-year period, while two years later President Clinton recommended that national R&D expenditures increase from 2.5 percent to 3 percent of national income. Meanwhile, the president of the National Academy of Sciences advocated a doubling of federal budgets for "fundamental research and training" within a decade, the Federation of American Societies for Experimental Biology called for an immediate 13.4% increase in research funding for the National Institutes of Health, and the chairman of the board of directors of the American Association for the Advancement of Science called for a "doubling of the [federal] research budget."[6]

The logic behind the argument seems apparent: if research and progress in science and technology are necessary for a better quality of life, then the more we spend on research the better our quality of life will be. This is the myth of infinite benefit. Researchers with impressive credentials in academia and private industry provide the legitimacy for this promise. Every year scientists, engineers, university administrators, corporate laboratory managers, and other members of the R&D community testify before congressional committees in an attempt to justify ever greater public expenditures for research. The myth of infinite benefit is explicit in their words: "The more innovation we have, the more competitive we will be as an economic entity, and the healthier we'll be as a nation."[7]

While a growing R&D system is portrayed as the key to the future prosperity of the nation—"Only through more research and more research can we provide the basis for an expanding economy"[8]—the most potent source of empirical and rhetorical support for this portrayal is the past:

> Once upon a time American science sheltered an Einstein, went to the moon, and gave the world the laser, the electronic computer, nylon, television, the cure for polio, and observations of our planet's location in an expanding universe. . . .
>
> America has lived and grown great through science and technology. From the founding of land grant universities and the flowering of agricultural research in the 19th century to the boom in microelectronics and information technology in the last two decades, we have hitched our economy to the best scientific research system we could develop and have prospered as a result.[9]

Thus did the scientific enterprise allow the nation to thrive in the past, and so does a bigger and better enterprise promise to nourish an even brighter future: "Science funding has increased steadily in the past several years, yet it is apparent that current levels are far below what is required for healthy, even lean science. . . . [A] sustained flourishing of academic research requires annual real growth of eight to ten percent."[10]

Recitation of past research breakthroughs demonstrates that science and technology are important components of a vibrant, growth-oriented society, an assertion that no responsible observer would deny. But understanding the value of a healthy research system is quite different from the expectation that an ever-larger system will yield ever-greater benefits for society. The distinction between these two points is rarely made, yet it is crucial because it determines whether the principal focus of R&D policy should be the size and cost of the system or the relationship between the structure of the system and the needs of society.

If one compares the specific words used to promote the myth of infinite benefit with the reality of societal evolution over the past fifty years, a rather startling pattern emerges. For example, Vannevar Bush's promise that more research would "mean more jobs, higher wages, shorter hours, more abundant crops, more leisure for recreation, for study, for learning how to live without the deadening drudgery which has been the burden of the common man for ages past"[11] could go uncontradicted for perhaps the first twenty years after World War II in part because the United States was the only major economic power to survive the war with its industrial base intact. Since the late 1960s, however, Americans have worked longer hours for lower wages, leisure time has decreased, concentration of wealth and the gap between wealthy and poor Americans has increased, the percentage of new jobs that are low paying and require a low level of skill (and, presumably, a high level of drudgery) has increased, unemployment rates have, on average, been higher, and overall indicators of social well-being have declined.[12]

Medical care in the United States offers a conspicuous contradiction to infinite benefit. The promise of more biomedical research delivering better health and better health care to Americans has been echoed from Vannevar Bush to this very day. Since 1970, no major segment of the national research budget has increased as rapidly as that for biomedical research. In the 1980s, research funding for medical sciences rose 41 percent *after* inflation; for basic biomedical research the increase was nearly 80 percent.[13] Even in the mid-1990s, when funding levels for most other areas of research were either static or in decline, support for biomedical research continued to grow. Despite these increases, the U.S. health care system is widely acknowledged to be in need of comprehensive reform as medical costs rise unabated (driven in part by the costs of new technologies and drugs), access to medical care increasingly becomes a privilege of those who can afford it, and public dissatisfaction with the system plays a growing role in national politics. International comparisons are particularly revealing. The United States government devotes more than 30 percent of its civilian R&D budget to health research, while other industrialized nations typically spend less than 5 percent. In absolute terms, the United States spends ten times more on health R&D than its nearest competitors, Japan and Germany. All the same, it lags behind other industrialized nations in many basic health indicators, ranking seventeenth in feto-infant mortality, nineteenth in maternal mortality, eighteenth in life expectancy from birth, and tenth in life expectancy from age 65.[14]

Even from a strictly economic perspective, the case for infinite benefit proves difficult to extrapolate from reality. While economists agree that the overall national investment in research yields a healthy rate of financial return, this conclusion does not imply that an increased investment will yield a larger return: "The anecdotal and historical evidence is . . . insufficient for advising on whether the current level of investments in science and technology is too large or too small."[15] Economic analyses stress that long-term problems such as the declining growth of industrial productivity and poor international economic competitiveness of some U.S. industries cannot be correlated with or attributed to national patterns of research spending.[16] Even in the case of highly profitable, research- and technology-driven industries such as pharmaceuticals, there is no clear empirical support for the myth of infinite benefit: "Whether a decrease in [pharmaceutical] R&D would be good or bad for the public interest is hard to judge. It is impossible to know whether today's level of . . . R&D is unquestionably worth its costs to society."[17]

The obvious rebuttal to these types of observations is that research cannot be blamed for the political or social or economic failings of the nation. Frank Press, former president of the National Academy of Sciences, explains, "What's really to blame . . . are things like the high cost of capital, the corporate culture with its short-term perspectives, tax policies that provide disincentives to do research, and leveraged buyouts. So I say, fix the system, don't put the blame on science, don't hurt the $15 billion research enterprise. Compared to other costs, it's dirt cheap."[18] Moreover, as a Nobel Prize–winning biologist argues, we have the scientific answers to many intractable problems; if they remain unsolved, the culprit is society, not research: "Science has long since produced the vaccines required to control most childhood infections in the United States, but our society has not found the political will to properly deploy those vaccines."[19]

These types of arguments have much superficial merit. They portray a world in which society and its problems are largely independent of the R&D system and its products. A nationally known physicist and leading voice for the scientific community observes:

> Environmental contamination at the nuclear weapons complex is a problem inherited from the past. . . . Then there is the savings and loan scandal—another problem of the past that we are paying for now. We are also trying to deal with the problems of our public schools, with an increasing homeless population, with deteriorating roads and bridges, with a rising rate of infant mortality. . . . What I miss in our country today is a sense of the future. We seem to be dealing with emergencies and paying for the years of neglect. In contrast, science looks to the future. If this country is to have a future, science must play an important role.[20]

In other words, if the nation is to overcome these problems "inherited from the past," it must look to science in the future. Implicit in this argument is the claim that the problems facing society today are unrelated to science and technology, whereas additional research will enable society to free itself from such problems tomorrow.

As a whole, the myth of infinite benefit isolates research, the purveyor of benefits, from the rest of society, the source of our ills. Yet if several decades of growing research and development expenditures have been accompanied by growing signs of societal stress in areas as diverse as economic performance, health care, and environmental quality, and if, as is widely acknowledged, the United States has the best research system in the world,[21] then perhaps the most benign conclusion to be drawn is that the absolute amount spent on research is irrelevant to society's ability to meet its most pressing challenges. As one industrial research leader explains: "Even at the national level, I can find reasons to question the belief that research inevitably enhances our economic well-being. While the US research investment might be comparable to that of other nations on a per capita basis, in total it is so massive that the results should be overwhelming. Yet there are many countries where the quality of life seems comparable or superior to that in the US. . . . [Quality of life] cannot therefore be linked to total research investments in [any] field."[22] The head of a major corporate research lab adds: "Trying to cure poor industrial performance with *more* university research is like getting helpers to push on a rope."[23] It is perhaps no coincidence that the foregoing observations come from scientists who work in the private sector and are thus not dependent on the government for R&D support.

Indeed, the benefits of an ever-expanding R&D system need not be understood purely in terms of societal well-being. Concealed within the myth of infinite benefit is political reality: a growing research system creates a growing research constituency and a growing demand for federal R&D funding. One of the purposes of the academic research system is to produce new scientists. Over an entire career, the typical science professor at a research-oriented university trains and releases into the world about twenty new Ph.D. scientists.[24] Of course, many of these young scientists go to work in the private sector; but many others stay in academia or work in government laboratories, where they require federal funding in order to conduct their research. This expanding population of scientists exerts continual stress on the federal budget process. During times of fiscal constraint, such as the budget crisis of the mid-1990s, the stress intensifies, and the result is a Malthusian struggle for survival. Competition for funding increases at every level: between scientists, disciplines, laboratories, universities, federal agencies, and congressional committees. Universities and other research institutions that have grown dependent on federal research support base tenure and promotion decisions for their faculty on productivity measured by the numbers of federal grants received, papers published, and graduate students supported. This, of course, stokes demand for funds still further. Meanwhile, graduate students themselves become postdoctoral researchers and faculty members, adding to the population whose work needs to be supported, and churning out more grant proposals, more publications, and more graduate students in the process.[25]

It is certainly no coincidence, for example, that biomedical research—the field that has received the most generous increases in government funding since the early 1980s—is also the source of the loudest and most insistent complaints about inadequate funding.[26] Indeed, fierce competition for funding among biomedical researchers and laboratories has risen more or less simultaneously with rapid growth of government support. Increased federal spending fuels, rather than quenches, this competition by providing support for more graduate students and faculty and by drawing more universities and laboratories into the research system.

In other words, the population dynamics of the R&D system stimulates an ever-increasing demand for government support—growth that can be rationalized by the myth of infinite benefit. This apparent confluence of the interests of the R&D community and the putative interests of society may be no more than coincidence, but there is some danger that the two may become confused. An example of this confusion recently played itself out around the issue of the scientific workforce. Throughout the 1980s and early 1990s, members of the research community and their advocates in government and industry argued that there was an impending shortage of scientists and engineers in the nation and that increased federal funding was needed to attract more students to scientific disciplines. Indeed, this viewpoint has not entirely died out.

The National Science Foundation, which funds research primarily at U.S. universities, issued several reports warning of a severe "shortfall" in scientists starting in the 1990s.[27] A 1990 report by the Association of American Universities stated: "Prompt action must be taken by all practitioners and patrons of doctoral education—government, industry, universities, and foundations. The *federal role* should be to provide increased incentives for talented U.S. students to enroll in doctoral programs and to assist universities in restoring the quality of the academic research environment . . . by providing expanded, flexible support for research and direct funding for research facilities and instrumentation."[28] Leaders of the research community took up the call. The president of the American Association for the Advancement of Science focused his 1990 presidential address on the shortfall "crisis."[29] Witnesses at congressional hearings highlighted the issue: "The impending shortage of scientific manpower is perhaps one of the most serious problems facing [the National Institutes of Health] and the biomedical research community in the 1990s," said one federal research administrator.[30] A respected academic researcher and administrator echoed this claim: "Simply stated, we face a severe and growing shortage of scientists and engineers. The statistics are frequently quoted but they bear repetition and remembering."[31] The specter of a shortage of scientists fueled a minor sort of hysteria. The press covered the story in terms that were almost apocalyptic: "With fewer students interested in science to begin with and a critical shortage of scientists and engineers, the attrition rate could soon represent a national crisis."[32] Congress responded to this threat by rapidly increasing funds for science and engineering education and by increasing the number of scientists and engineers allowed to immigrate to the United States from other nations.[33]

Meanwhile, many scientists who were in the process of receiving their Ph.D.'s or who had recently completed their studies were discovering a disturbing fact: good jobs were increasingly hard to come by in many fields. Competition for tenure-track positions at universities was fierce; in many disciplines hundreds of applicants found themselves vying for a single position. At the same time, many sectors of private industry were downsizing their research laboratories.[34] A disillusioned physicist founded an organization called the Young Scientists' Network whose explicit goal was "[to] let the press, public, and government officials know that there is no shortage of scientists. Instead there is a glut of scientists."[35] By 1992, NSF's projections of a shortfall of scientists were discredited and repudiated, and the "crisis" had waned. The tone of the press coverage changed abruptly. The lead paragraph of a 1992 *New York Times* article read: "Even as leading scientists warn that America's education system is failing to produce scientists fast enough to fill a glaring projected shortage, many young physicists contend that universities are already turning out far more physicists than there are permanent jobs."[36]

But before the scientist shortfall scare had abated, Congress had been manipulated into providing more funds for the research system and skewing national policies toward increasing the research workforce. There is no need to look for a conspiracy here.[37] The universal acceptance of the myth of infinite benefit caused scientists and policy makers alike to embrace uncritically a call for growth in the research system that contradicted common sense and was unsubstantiated by either good data or experience. The NSF shortfall predictions were based on unsound analytical methodology, and they were never subjected to the rigorous outside peer review that the research community cites as the foundation of its own credibility.[38] This community showed no reluctance in moving aggressively into the political arena while trumpeting flawed work that violated its own standards of excellence and accountability but advanced its own agenda for growth. Early criticisms of the shortfall predictions were ignored, perhaps because they came from social scientists and engineers,[39] who are not accorded the same status as the natural scientists who dominate the leadership of the research community and who led the hue and cry. But perhaps the strangest aspect of the whole shortfall "crisis" was that it occurred at a time when growth in the number of researchers in the United States had greatly increased competition for federal research funds. Thus, at the same time that scientists were arguing that there was an impending shortage of researchers, they were also clamoring for more federal funding to offset the effects of increased competition. Robert M. White, president of the National Academy of Engineering, explained the origins of this contradiction: "The nation has loosed a flood of scientists and engineers in response to real and perceived national needs. Increases in government and private sector spending, generous by any standard, have brought total R&D funding to $150 billion in 1990, but have not been able to sustain the increases in the R&D community's population."[40]

As the size and cost of the research system has grown over the past half century, so has its dependence on public opinion, public funding, and political influence. Under such circumstances, the myth of infinite benefit is a potent political weapon. It confers sufficient power and legitimacy on the research community to influence public policy and the expenditure of public funds and may therefore be an effective catalyst for self-interested action, as illustrated by the shortfall "crisis." NSF's shortfall predictions were based on the assumption that society's demand for scientists would rise smoothly and predictably into the indefinite future. For many members of the R&D community, the possibility that market forces and government policy would fail to stimulate perpetual growth of the research enterprise was unimaginable, even unconscionable. One scientist went so far as to publicly characterize his inability to attract sufficient federal funds to support all of his graduate students as "federal suppression of science,"[41] as if the direction of social and political obligation ran from society to science, rather than vice versa. Overall, the shortfall predictions ignored the fact that the research system was an integrated part of a larger society and was therefore subject to economic, political, and social pressures and constraints.

Herein lies the greatest problem. Based, as it is, upon promises for the future and anecdotes about the past, the myth of infinite benefit obviates consideration of the role of science and technology in society in the present. If the flow of benefits from laboratory to society is automatic, then policy decisions largely devolve to questions of "how much" and "how many." A "crisis" means that there isn't enough money for all qualified scientists to pursue all interesting questions, or that the number of scientists will not continue to grow at the same rate as it did in the past.[42] A more realistic perspective, however, would include the context of the society that assimilates and absorbs the products of the R&D system:

> We [scientists] shouldn't create the illusion that science alone brings these enormous benefits. [Solid state] physics alone did not create the computer industry, though there wouldn't be a computer industry without solid-state physics. We have to develop a convincing and accurate picture of the role of science in our society, that it is part of a complex process that ends up with products and prosperity. We need to instill a realistic picture of cause and effect, not a picture that science does it all. [Forty years ago] a career in science . . . did not depend on ever-accelerating growth in the field. There was a more or less stable number of scientists. . . . After [World War II] there was a stream of money from the government, so people's eyes turned naturally in that direction, and scientists became divorced from the rest of society. . . . I believe the absence of perpetual growth will be healthy for science. It forces us to confront our true role in society.[43]

These are the words of a physicist, and they suggest that alternatives to the myth of infinite benefit are not intrinsically "antiscientific" even if they are not entirely congruent with the short-term political and professional interests of the federally funded R&D community. Rational, forward-looking R&D policy cannot be designed and implemented if the "true role" of science and technology in society is poorly understood. This understanding is not implicit in scientific or technological expertise but requires a more distant perspective that views the laboratory as an element of the surrounding society. At issue are fundamental questions about the relation between research and the fabric of modern culture, questions that may appear hopelessly vague in part because there have been so few serious attempts to answer them (or to find ways to ask them more precisely). How can we begin to evaluate the balance between benefits reaped by science and problems imposed? Does this balance depend in any way on the research itself or on the structure of the research system? What types of societal problems are likely to be responsive to more research and what types have proven relatively insensitive to scientific and technological advance? What types of research have proven most effective in addressing social problems? How are the long-term goals of a democratic society best served by research? In light of ongoing fiscal constraints, how large should the R&D system be, and how should finite resources be allocated, in order to make the best use of available funds and ensure the best return to society on its investment?

Until researchers and policy makers alike begin to address such questions, there can be no rigorous basis for arguing, for example, that continued increases in the research budget for the National Institutes of Health will translate into higher levels of public health, or that overall reductions in federal R&D expenditures—such as those pursued by Congress in the mid-1990s—will have a negative societal impact. One need only consider the fate of nuclear energy in America—extravagantly funded by the government, passionately embraced by the utilities industry, resoundingly rejected by the public—to appreciate the distinction between the promise of infinite benefit and the effects of science and technology on a complex democratic society. Current federal R&D initiatives, such as programs to map the human genome and to establish a nationwide high-speed computer network, are similarly destined to have significant unanticipated impacts on society that may compromise their anticipated contributions to quality of life. As long ago as 1947, the Steelman Report to President Truman included the following quotation from a representative of the National Research Council: "I cannot think of any field of research in physical science which does not ultimately lead, and usually very promptly, to new social problems. The same is true in biology and medicine. It is important, therefore, that competent social scientists should work hand in hand with the natural scientists, so that problems may be solved as they arise, and so that many of them may not arise in the first instance."[44]

There has been little substantive effort to implement such a vision. On the whole, natural and social scientists still tend to regard each other with suspicion or hostility—to the extent that they regard each other at all[45]— and few formal mechanisms exist to promote cooperation between the two. The potential value of such an alliance will be discussed in the final chapter. The more significant point here is that the myth of infinite benefit justifies the research community's single-minded attention to the size and cost of the R&D system and thus distracts from the underlying— and much more difficult—policy problem of how to create the greatest possible compatibility between the new knowledge that scientists create and the capacity of society to assimilate this knowledge for its long-term benefit—regardless of absolute levels of expenditure. One way that leaders of the R&D community evade consideration of this problem is by arguing that the creation of explicit linkages between a given research program and a particular area of societal need is often impossible. They base this assertion on a complementary myth—the myth of unfettered research—which states that the results of many types of research are intrinsically unpredictable and that the benefits that may accrue from such research must therefore be equally unpredictable.

The Myth of Unfettered Research

[How] different everything would be if we in the Orient had developed our own science. Suppose for instance that we had developed our own physics and chemistry: would not the techniques and industries based on them have taken a different form, would not our myriads of everyday gadgets, our medicines, the products of our industrial art—would they not have suited our national temper better than they do? In fact our conception of physics itself, and even the principles of chemistry, would probably differ from that of Westerners; and the facts that we are now taught concerning the nature and function of light, electricity, and atoms might well have presented themselves in different form.

Of course I am only indulging in idle speculation; of scientific matters I know nothing.—Junichiro Tanizaki, *In Praise of Shadows*

A LONE SCIENTIST, frizzy-haired and bespectacled—the absent-minded, benevolent genius lost in thought; or perhaps the dedicated experimentalist clad in a white coat and laboring madly among the condensers, Van de Graaff generators, computers, and even electrode-covered cadavers: These are typical public images of scientific research. Real scientists protest such portrayals as distorted caricatures and sources of public misunderstanding. But the research community itself promotes an image of the scientist that is not so very different—that of the intellectual maverick isolated from the hubbub of daily life, restricted by nothing but the limits of imagination, dedicated to exploring the frontiers of knowledge.

"[Three] intelligent and well-educated young men were tested for their wisdom, diligence and ingenuity by their father, the king, in the following way: They were to travel across the world in three different directions with the mission of finding a potion to kill the fearsome dragons that were surrounding and isolating the island of Serendip." So begins a promotional pamphlet from the American Chemical Society entitled *Science and Serendipity: The Importance of Basic Research.* It continues:

> The princes never found the magic formula. . . . But in the course of their travels they solved mysteries, saved a nation from starvation, slew a magic serpent, restored property to its rightful owners, saved a king from poisoning, and rescued a lost princess. . . . When they returned to the King, even though they had not achieved their original purpose, he realized that they were wise enough to rule the Kingdom. . . .
>
> Many people believe—having read about the life of Thomas Edison—that useful products are the result of targeted research, that is, of research specifically designed to produce a desired product. But . . . progress is often made in a different way. . . . Like the princes of Serendip, researchers often find different, sometimes greater, riches than the ones they are seeking.[1]

Today, the autonomous and independent scientist, questing after knowledge, is supported not by a king but by the federal government. Basic research—the investigation of natural phenomena (also called fundamental, pure, and curiosity-driven research)—does not offer the promise of immediate or short-term economic payback to its patrons. Thus, a core precept of science and technology policy in the United States is that the motivation and responsibility for funding basic research lies primarily with the state rather than with private industry. For more than forty years, the U.S. government has supported the majority of the nation's basic research, and expenditures for this activity have risen progressively, from about $250 million in 1953 to almost $14 billion in 1995, an eight-fold increase after inflation.[2]

As with so many other aspects of postwar science policy, perhaps the most lucid and influential justification for a federal role in supporting basic research was articulated by Vannevar Bush in *Science, the Endless Frontier*:

Basic research is performed without thought of practical ends. It results in general knowledge and an understanding of nature and its laws. This general knowledge provides the means of answering a large number of important practical problems. . . .

One of the peculiarities of basic science is the variety of paths which lead to productive advance. Many of the most important discoveries have come as a result of experiments undertaken with very different purposes in mind. . . .

Basic research leads to new knowledge. It provides scientific capital. It creates the fund from which the practical applications of knowledge must be drawn.[3]

From a policy perspective, the importance of this portrayal of basic research is that practical benefits accrue to society through an apparently unrelated process—the creation of a reservoir of knowledge about the structure of nature. Because this structure is not known in advance, it is not possible to predict either the direction of scientific progress or the specific practical benefits that may result from this progress. Therefore, efforts to steer basic research in any particular direction are allegedly doomed to failure. Because the motivation for the research is curiosity about nature rather than the resolution of practical problems, researchers must be shielded from political, economic, and social pressures and restricted by nothing but their own abilities and imaginations.

The organization of the federal basic research enterprise generally reflects these considerations. The academic environment is often viewed as ideally suited for basic research, and about half of federal basic research funds—$7 billion in 1993—are awarded to universities.[4] Agencies such as the National Science Foundation and National Institutes of Health disseminate basic research funds to scientists—the majority of whom work at universities—based on criteria of scientific excellence and using evaluation mechanisms such as peer review of research proposals. In this way, funding decisions are largely protected from both external politics and expectations of practical application, while researchers are granted a large degree of autonomy in pursuing their investigations. What are the promised benefits of this system? A Nobel laureate in medicine explains: "The pursuit of curiosity about the basic facts of nature has proven throughout

33

the history of medical science to be the most practical, the most cost-effective route to successful drugs and devices. Investigations that had no practical objective have yielded most of the major discoveries of medicine—X-rays and penicillin, the polio vaccine, monoclonal antibodies, genetic engineering, and recombinant DNA. These have all come from the pursuit of curiosity about questions in physics, chemistry, and biology, apparently unrelated at the outset to a specific medical problem."[5]

Whereas the myth of infinite benefit suggests that more science and technology leads to more societal benefit, the myth of unfettered research dictates that any line of basic research is as likely to lead to societal benefit as any other. Just as no one foresaw the electronics revolution prior to the discovery of the transistor effect, and no one anticipated the biotechnology revolution prior to the determination of the structure of DNA, neither can anyone know what particular area of investigation will lead to future scientific and technological revolutions. Because of this impossibility, the myth implies that the most efficient approach to basic research is to maximize the production of new knowledge—any new knowledge—and thus maximize the likelihood of serendipitous and revolutionary discovery. A leader of the biomedical community makes the case for his discipline:

> [New] technologies are rooted in a science base, of course; and most of that knowledge in turn derives from investigator-initiated research, especially those inquiries that probe the most fundamental functions and forms of the living state. To many, this fact is self-evident and therefore a prima facie justification for continuing to spend public and private dollars on self-starting scientists.
>
> Others, however, are skeptical about expecting a collection of asynchronous, independent initiatives to achieve particular societal goals. They note that basic research by its essence is not an intuitively obvious route to any preconceived destination. They are prone to expect more utility from finely focused projects than from free-ranging studies of natural phenomena. They are more comfortable with consensus than with idiosyncrasy. They favor road maps over serendipity. And they find disquieting the apparent paradox that, in science, the one who aims most intently at the target often is the one least likely to hit the bull's eye![6]

A leading physicist presents a similar argument: "In the U.S. most basic research in the physical sciences is carried out by small groups at universities. . . . They are free to move rapidly in new directions, and they have for some time been the mainstay of science in the U.S. Because the research is basic, the groups are guided by scientific curiosity and propelled by vision and imagination rather than the need to solve a practical problem. Yet practical application is often not far behind."[7]

Such portrayals suggest that the capacity of the basic research system to contribute to societal welfare depends on a single independent variable: the scientist in the laboratory, "free to pursue the truth wherever it may lead."[8] So long as scientists are unconstrained by politics or by thoughts of practical application, and so long as they have the resources necessary to pursue their curiosity in uncovering the secrets of nature, knowledge will expand and, it is said, society will benefit.

This model of progress thus treats fundamental scientific knowledge as an organic outgrowth of the interaction between curious scientists (assisted, of course, by whatever experimental apparatus they may chose to employ) and undiscovered natural phenomena—nothing more. The model defines an ideal—the unfettered researcher—and equates the ideal with the entire basic research system. According to this equation, the research system should be designed so that the basic scientist is constrained by nothing but the inherent organization of nature itself. And there can be no question that, on the scale of the individual, a scientist may indeed be free to pursue some intriguing avenue of inquiry. But that scientist is also part of a larger research system, and the organization of that system is a consequence of a complex set of still larger political, economic, historical, social, and scientific interactions. From a policy perspective, therefore, a scientist is an almost incidental element of this broader context, and a research problem reflects not just the intimate relationship between scientist and nature but the evolution of the research system as a component of society.[9]

External Fetters: Teapot in a Tempest

Suppose there were such a thing as an ideally unfettered research system, subject to revolutionary, serendipitous breakthroughs, creating new knowledge in a manner entirely unpredictable yet inexorable, driven forward only by the interests and abilities of the researchers. The only controls on such a system would be the ability of the individual scientists, the resources available to them, and the existing state of scientific knowledge. There would be no way to judge the ultimate value to society of any particular line of research or even to guess the most likely type of contribution that a line of research would make. Therefore, the most sensible approach to science policy would be to divide funds up equally among the best researchers in the various disciplines.

In fact, such a course of action was suggested by a distinguished physicist testifying before Congress:

> If I were King, I could take the budget of the National Science Foundation and spend it with ten times the effectiveness, and I could do this with little effort. I would find the top one thousand active scientists in the country. I could do this by asking all the scientists for their list of the best, and then asking these scientists for their list of the best. By the next iteration, I believe that most lists would agree. Next, I would give each of these 1000 scientists one million dollars per year to spend on whatever research they wanted. The money spent this way would, I believe, be far, far more productive in producing scientific innovation than by doling it out for proposals that achieved the consensus of [peer review].[10]

Few basic researchers would endorse such a plan. Biomedical scientists lobby Congress for increased funding in basic biomedical research, not in high energy physics or polymer chemistry. They argue that increased federal funds in their own discipline will ultimately make a contribution to medical treatment and technology, not to the development of new energy sources or plastics.[11] Although basic research is portrayed by scientific leaders as essentially unpredictable and serendipitous, as a practical matter policy makers and researchers appear to believe that the pursuit of fundamental understanding in a given field will give rise to applications primarily in that same field. It is no accident that, while basic biomedical research today receives more federal funding than any other basic science discipline, applied biomedical research is the most generously supported of all applied disciplines. Nor is it an accident that physics is second in both categories, or that anthropology is second to last among basic sciences and last among the applied sciences.[12] Overall, the amount of federal funding devoted to a given area of basic research is a strong indication of the level of political and economic interest in the anticipated outcome of that research.* The number of scientists pursuing their unfettered curiosity in a field is testimony not to some level of intrinsic intellectual worthiness but to the availability of federal dollars.†

Opportunities for progress in basic research may also depend on important problems or phenomena revealed by applied research and technological innovation. This has been the case since the beginning of the industrial revolution. For example, in the early nineteenth century the invention of the steam engine helped to stimulate basic research in thermodynamics, and research into the properties of metals was stimulated by technological advances in steelmaking. Fundamental research in radio astronomy grew out of the efforts to improve overseas telephone service in the 1930s. During World War II, the search for a way to track German U-boats led to magnetic mapping of the sea floor, an essential foundation for the plate tectonics revolution in basic geological sciences. Applied investigations into the causes of pneumonia led inadvertently to the discovery of DNA, which spawned entire new fields of basic biological and medical research and applications. Today, the $500 million-a-year U.S. government investment in basic research on global climate change is motivated by 150 years of technological innovation and consequent increased emission of greenhouse gases. In cases such as these, technology is the stimulus for basic research, not the consequence of it. The relation between fundamental research and application is one of complex interdependence. Each may drive the other forward.[13]

Nor is serendipity the exclusive domain of basic research. In fact,

---

*A few basic research fields have little potential practical application but still receive significant funding for political reasons. Robust federal support for astronomy, for example, is rooted primarily in federal expenditures on the space program, which in turn grew out of the Cold War and competition with the Soviet Union for technological supremacy in space.
†Funding agencies may choose to focus more resources on particular "hot" areas of basic research that seem ripe for rapid advance (even if they have no known application), but the scale of such decisions is small relative to the basic research budget of any major program, field, or federal agency.

unanticipated discoveries and innovations often emerge from the applied research laboratory. Teflon was an inadvertent product of applied research on refrigerants, the microwave oven was an accidental outgrowth of research on radar, and mass production of penicillin was made possible because government researchers were trying to find a way to get rid of agricultural byproducts that just happened to encourage growth of the penicillin mold.[14] Any creative researcher may take advantage of unanticipated experimental results to make surprising discoveries and inventions, regardless of whether the science is "basic" or "applied."

Furthermore, high quality basic research *can* be carried out with a view toward potential applications. The Manhattan Project included a significant component of basic research in nuclear physics which was motivated solely by the need to design the first atomic weapons. The experience of the Manhattan Project shows that basic researchers who are forced to think in terms of applications may be spectacularly successful in forging linkages between fundamental knowledge and the use of that knowledge by society.[15] Similarly, it is no coincidence that many of this century's most revolutionary technological innovations—including the transistor, the laser, and nylon—were the consequences of basic research carried out in whole or in part at industrial laboratories such as Bell Labs and Dupont. Although these laboratories gave their researchers freedom to pursue their curiosity, they also imbued in them an underlying responsibility to contribute to the long-term goals of the company. "Even in the ivory tower of Bell Labs in the 1960s," said one researcher, "we never forgot the fact that we existed to serve the telephone business."[16] Overall, a sensitivity toward potential applications appears particularly conducive to promoting useful discoveries from basic research. There appears to be no reason why research carried out in the general context of a particular application should be less likely to achieve some unexpectedly useful result than research that is totally isolated from practical goals. Indeed, scientists who are uninterested in applications may be less willing to search for them and less able to recognize them.[17] According to one of the discoverers of the laser: "For this investigation, we needed no reason other than the history of radio. We knew that whenever people had found ways to generate shorter wavelengths, *those wavelengths had been found useful*."[18] As well, part of the great success of both Bell and Dupont in translating advances in basic research into important practical innovations has been attributed to an intellectual environment in the laboratory that encouraged interdisciplinary research and open communication between basic and applied scientists at all levels and in different fields.[19]

Political and historical milieus strongly influence the course of basic research. Scientists are preferentially able to satisfy their curiosity—to "pursue the truth wherever it might lead"—in fields that are closely linked to political and economic priorities. Physics has been such a field until recently, and for obvious reasons: since World War II, physicists have been among the most influential science policy voices, and they owe this status to the success of the Manhattan Project and the importance of nuclear weaponry in the Cold War. As one commentator observed: "It's not just that physicists were smarter or more charismatic than other scientists—or even that they gave the world the transistor. They had built the Bomb. As the Merlins of the Cold War, their wizardry could tip the balance of superpowers in the twinkling of a quark."[20] Funding for basic research in physics has been intimately coupled to the nation's defense and weapons development agenda. A federal commitment to civilian nuclear power following World War II also created a significant economic constituency for basic research in physics.

Inevitably, the disciplines that attract the most funding are also those that have the most political and economic clout. These are mutually reinforcing phenomena. For forty years, the interdependence of physics and the Cold War helped justify increased budgets for basic research in physics, and these increases in turn meant more influence with Congress, with funding agencies, and with universities. Of course such relationships are further strengthened by the progress of science. A well-funded discipline will have many more opportunities to advance the state of knowledge than a poorly funded one, even if both have a potentially important contribution to make to society.

Since 1970, federal funding for biomedical science has grown much faster than for other fields. These increases have translated into more political power for biomedical researchers, both in absolute terms and relative to their colleagues in other disciplines. Today, the interdependence of basic biomedical research, the medical industry (including the pharmaceuticals, biotechnology, and medical technologies industries, as well as the health care delivery system), interest groups who lobby for more research on specific diseases, and a voting public greatly concerned about health issues in general creates a constituency for medical research as powerful as the physics constituency during the Cold War. Such trends have considerable inertia; they are fetters hardened by history, economics, and culture. They confine the pursuit of knowledge to particular paths. Even if it were shown today that basic biomedical research was failing to make a cost-effective contribution to society—or that alternative fields had the potential to make a significantly greater contribution— funding levels for biomedical science would continue to dominate the federal research budget for years to come because of the political and economic strength of the biomedical research community. The scientific dominance of a particular field of basic research arises in part from historical trends that are not easily changed.

One might argue that the dominance of biomedical research must be traced not to political and economic forces but to the revolution in molecular biology that has catalyzed entire industries and offered hope to those afflicted with debilitating disease. But this argument does not address the fact that other industrialized nations with the scientific capability to exploit this revolution choose to spend considerably smaller proportions of their research budgets on biomedical research than does the United States. The dominance of biomedical fields over all other areas of research in the United States is not the result of an inevitable intellectual evolution that began with the discovery of the structure of DNA. It is a choice made by society, imposed in a manner that is often more ad hoc and capricious than strategic, and fueled in large part by government spending.

Not surprisingly, then, when basic research disciplines lack a powerful political or economic constituency, they are also lightly funded. In the case of environmental science, for example, the slashing of an already modest basic research budget at the Environmental Protection Agency (EPA) in the early 1980s was greeted with a roaring silence outside of the environmental community. Basic research at EPA found little support in the growth-oriented, supply-side economic philosophy of the Reagan administration. Neither did it have much of a constituency within the scientific community, probably because it was a small, nonmainstream program carried out mostly by government scientists and aimed at specific societal problems.[21]

Basic research in the social sciences has also received little support from the government—typically about one percent of the total basic research budget.[22] This neglect in part reflects the attitude of leaders of the science policy community, who often hold the social sciences in low regard. Vannevar Bush was notably hostile to the social sciences, and his original conception for the National Science Foundation excluded them entirely. Furthermore, Bush and his supporters—many of whom were academic physical scientists—feared that inclusion of the social sciences in their plan for NSF would create opposition in Congress, thus threatening the prospects for federal support for their disciplines as well. Historian of science Daniel Kevles highlighted Bush's position: "Any provision for the support of the social sciences would be 'dynamite.' Social scientists could not easily convince Congress to spend federal funds 'for studies designed to alleviate [the] conditions of the Negroes in the South or to ascertain the influence of the Catholic Church on the political situation in Massachusetts.' "[23] Thus, meager funding for social sciences is not attributable to a dearth of interesting or significant research problems, nor to a lack of potential contributions to society—especially if such contributions are supposed to be intrinsically unpredictable and serendipitous.[24] Rather, it is a product of factors extrinsic to the research process, including social and intellectual hierarchies within the research community and the ideological and economic motivations that underlie federal support for science.

There is, in essence, a basic research market, driven not only by the curiosity of talented scientists and the work of their predecessors but also by funding levels, job opportunities, public expectations, economic interests, and politics. The knowledge produced by basic researchers is an indirect reflection of the priorities of this market. The politics and economics of the research system are as subject to the whims of serendipity as the research itself. The ideological struggle between East and West, and the fear of unanticipated scientific or technological breakthroughs on the part of the enemy, were the primary motivations for government support of basic research from 1945 to 1990. Thus, while the entire world was held hostage to an arms race that could have led to the annihilation of humankind, scientific knowledge advanced at breakneck speed, riding on the back of ideological confrontation. Much of the knowledge created by scientists over the past fifty years therefore exists only because of the Cold War.[25] From this perspective, the dominant position of physics and physicists in the United States in the decades after World War II had as much to do with Stalin, Hitler, and Hirohito as it did with Einstein. This dominance was an accident of history reinforced by decisions of policy. As a result of this history and policy, basic research in physics has consistently received—and continues to receive—more funding than any broad discipline except basic biomedical research. But a different set of historical circumstances, different political and economic imperatives, and different social expectations might well have placed a higher priority on other types of research, leading to different discoveries and societal impacts.

Internal Fetters: The Maleness of the System

Directions of basic research are also influenced by internal characteristics of the R&D system that may seem at first to have little connection to the actual pursuit of scientific knowledge. The overwhelming "maleness" of the R&D community is one such characteristic. In the 1960s, almost 90 percent of the Ph.D.'s in science and engineering fields were awarded to men. By 1991 this figure had fallen to about 71 percent. The total proportion of women in the Ph.D.-level research workforce—in academia, private industry, and government—was about 16 percent in 1993, with female scientists strongly concentrated in biological, behavioral, and social sciences. In the physical sciences and engineering women still constitute considerably less than 10 percent of the Ph.D.-level workforce.[26] About 4 percent of the elected members of the National Academy of Sciences are women.[27]

So what? If science is merely the pursuit of knowledge about objective phenomena, and if the scientific method and peer review act to filter out the effects of individual personality and bias, why should the dominantly male composition of the research workforce have any impact on the progress of science? Isn't the question of "maleness" in the laboratory one of social equity rather than the conduct of research? As one physicist remarked: "Newton's laws work, and that's the only thing I need to know about Newton."[28]

The scientific validity of a given equation or hypothesis does not, of course, depend on the gender of the investigating scientist. But the research system that exists today was designed by and for men, and therefore it is men who have established its operations, priorities, standards, and objectives. Men have overwhelmingly made the decisions that determine which equations are solved and which hypotheses are tested. Would a research system in which men and women were equally represented, or a research system predominantly controlled by women, be identical to what we have now? Would it have the same research priorities and the same internal organization? Would it create the same knowledge and make the same discoveries?

Biomedical research offers the most straightforward perspective on such questions. Starting in the mid-1980s, evidence began to surface that the U.S. agenda for health research included numerous internal sources of bias against women. On one hand, basic and applied research on a range of women's diseases had received relatively low priority among biomedical scientists, while on the other, knowledge about a wide range of non-sex-specific diseases derived predominantly from studies of men. The most conspicuous and highly publicized problem lay in major clinical trials funded by the National Institutes of Health (NIH), including studies on the effects of caffeine on heart disease and the relation between aspirin consumption, heart disease, and stroke. These and many other trials included only men in the study population, even though the results of the trials were intended to provide improved medical treatment protocols for both men and women.[29]

The exclusion of women from clinical trials was justified along several lines, including the complicating effect of women's hormonal cycles on study results, the greater life expectancy of women (which would require longer and therefore more costly studies), and the risk of fetal damage for pregnant women.[30] This logic is baffling in its circularity. If the presence of women compromises the validity or the safety of a clinical trial, how can the results of the trial be validly and safely applied to the female population at large? Furthermore, the idea that hormonal cycles are a scientific inconvenience rather than a physiological reality made sense only because the leadership of the biomedical research community had been dominated by the gender that did not experience such cycles.

43

The impetus for change came, not surprisingly, from women outside the community, including two politicians, Representatives Olympia Snowe and Patricia Schroeder, cochairs of the Congressional Caucus on Women's Issues. They initiated the congressional investigations that provided much of the political and scientific incentives to create the Office of Research on Women's Health at NIH in 1990 and the $625 million Women's Health Initiative three years later.[31] Bernadine Healy—appointed first female director of NIH in 1991—explained the need for the initiative in terms that no male would have adopted: "For far too long research on women's health has been neglected. Men were the normative standard for medical research and treatment. The corollary for this, of course, is that men's hormones set the standard for us all."[32]

That a biomedical research agenda rooted in the norms of male physiology would evolve from a male-dominated biomedical research community is perhaps predictable. A more difficult question, however, is whether the historical dominance of men in the research system has had a more general impact on the pursuit of basic knowledge. Some have suggested that the behavior of women scientists tends to be organizationally different from that of men; that women, in contrast to men, favor collaboration over individuality, cooperation over competition, intellectual achievement over professional advancement, organizational success over individual status. In fact, several surveys lend preliminary support to these assertions, although many researchers, both male and female, are wary of this type of generalization.[33] Such arguments need not imply, however, that each individual man and woman brings a distinctive, inflexibly "male" or "female" research style into the laboratory; rather, they suggest that the prevailing research methodologies and agendas of a male-dominated research system reflect the biological and sociological influence of maleness and largely exclude the influence of femaleness.

Because males have so dominated science—especially the leadership of the R&D community—evidence bearing on this question is necessarily scarce, but the experience of Nobel Prize–winning biologist Barbara McClintock may be illustrative. In the 1940s and 1950s, McClintock was one of the few women occupying the upper echelons of the research hierarchy in biology. Her studies of mutation in corn, however, went very much against the grain of cutting-edge biological research. Most leading biologists, perhaps emulating the reductionist successes of physicists earlier in the century, were focusing their attention on the molecular structure of chromosomes and especially the structure and function of DNA in bacteria. The goal of such work was to understand how genetic characteristics were passed between generations, and this problem was apparently solved when the double-helix geometry of DNA was discovered in 1953. Thus, biology progressively became more mechanistic, more deterministic, more rooted in the elucidation of the structure of DNA, the "master molecule" from which all the characteristics of life derived. McClintock worked against this backdrop. Depending primarily on her own visual observations of color patterns in corn, she developed a theory of genetic mobility, or transposition, which dictated not only that the structure of a chromosome was intrinsically unstable but that certain breaks in this structure were controlled by genetic elements within the chromosome itself, responding to stimuli in the surrounding environment—that is, genetic mobility was intrinsic to the chromosome. (Decades later, this principle was used to explain how some bacteria develop resistance to antibiotics.) Completely out of phase with intellectual and methodological trends in molecular biology, her work remained misunderstood and unappreciated for almost thirty years.[34]

One can only speculate on the degree to which McClintock's research reflected the "femaleness" of her approach; certainly male scientists have made important discoveries that went unacknowledged because they were incompatible with the scientific trends of their times. (McClintock's intellectual forebear, Gregor Mendel, is one example.) McClintock's biographer, historian and philosopher of science Evelyn Fox Keller, highlights the conspicuous stylistic contrast between McClintock's work and that of most other leading (and male) biologists at the time.[35] While the scientific mainstream focused on determining rigid structure at the molecular level, McClintock was concerned with divergence from rigidity and control—with inherent instabilities that were visible to the naked eye. McClintock's own explanations of her methods focus on the importance of "seeing"—of the relationship between her mental state and her powers of observation—with an almost mystical fervor that some might describe as patently unscientific. She possessed an apparently visceral empathy for her samples to complement her intellectual rigor.[36] Again, such attributes are not unique to McClintock nor to women, yet neither are they representative of the norms of the scientific endeavor as established and enforced primarily by men.[37]

The possibility that women might bring a different collective experience into the laboratory implies at the very least that the existing ethos in the basic research establishment is neither inevitable nor immutable nor even necessarily optimal. A male-dominated research system has emphasized disciplinary research over interdisciplinary investigations; reductionist approaches over synthetic ones; natural sciences over social and behavioral sciences; individuality over collaboration. Such preferences affect the course of research and the advance of knowledge, but they are not intrinsic to science. As Evelyn Fox Keller argues, while the objective structure of nature is a principal control on the direction of scientific progress, it is not the only control: "Social, psychological, and political norms are inescapable, and they too influence the questions we ask, the methods we choose, the explanations we find satisfying, and even the data we deem worthy of recording."[38]

Although there is controversy over the possibility that women and men might bring different approaches to research into the laboratory, there is at least one undisputed difference between the sexes that has had an unmistakable impact on the participation of women in research: childbearing. This impact has been particularly amplified by the tenure system, a fundamental job-security provision of academic research that discriminates, in its essence, against women because they are women. On the face of it, this discrimination can again be dismissed as an issue of social equity rather than a constraint on research, yet it is intriguing to consider not merely the manner in which tenure may have restricted women's participation in university research but the more general effect of tenure on the character of the research system.

The price of tenure is usually a single-minded devotion to research during the first six or seven years of a scientist's professional career. This is particularly true at the most prestigious and influential research-oriented universities where the leaders of the research community are concentrated and where childbearing and research can become mutually exclusive. As one critic notes, "A woman is usually 30 years of age before assuming an assistant professorship at a university, which puts her tenure decision at age 35 to 36. Thus her critical scientific years, in which she is establishing her reputation, and her peak reproductive years coincide. This is a dirty trick. . . . Tenure is no friend to women. It does not protect them from institutional discrimination. Rather it rigidifies their career path when they need maximum flexibility."[39]

If women were full partners in academic research, perhaps tenure would no longer exist in university science and engineering departments.* Such a change could have significant positive consequences for the university research system, including greater motivation for older faculty to remain active; less job protection for unproductive, "deadwood" faculty members; more opportunities for young researchers to enter the system and fewer disincentives for them to pursue high-risk research problems; and perhaps an increasingly vibrant and dynamic academic research system on the whole. Other consequences might be negative. Perhaps quantitative measures of performance (such as number of publications or grants) would become even more important criteria for success in the absence of a lifetime job guarantee.

The intent of this discussion is not to argue that more women—and more influential women—in the research system would necessarily produce better science (although that possibility does exist). Rather, it is to suggest that more women would produce *different* science and such differences might ultimately address important scientific questions that would otherwise remain unexplored. Differences could arise from three sources: priorities and interests of women scientists that may not be fully shared by men; approaches to the conduct of research that may be more typical among women than men; and the possibility of structural

---

*Tenure was created to protect academic freedom; the relevance of this justification to most scientific and engineering fields today is debatable, but in any case it would be desirable and possible to implement safeguards against political abuse of academic hiring and firing in the absence of tenure. One rather whimsical alternative to tenure, suggested to me by a Columbia University research administrator, would turn the system on its head: professors would start out with a seven-year, no-questions-asked contract during which they would be free to establish their research and teaching programs without worrying about the quantity of publications and grants required for tenure, after which a series of shorter contracts would be signed based on the quality of the professors' work as their careers advanced. Such a system could, for example, give maximum flexibility to women who choose both to do research and to start a family.

changes—such as elimination or modification of tenure at universities—
that could influence the internal organization of the research system and
in turn affect the conduct of research by males and females alike. The
"maleness" of U.S. research should therefore be viewed as an intrinsic
fetter on the system that can influence choices, priorities, directions, and
goals of even the most basic research.*

Unfettered Reality

Curiosity-driven basic research is not carried out in isolation from the
rest of society. Of course serendipity and contingency are important parts
of scientific progress, just as they are in virtually all aspects of the human
experience. And of course scientists are uniquely qualified to determine
how a given line of research should be pursued in the laboratory. But
serendipity and the expectation of intellectual autonomy cannot liberate
the research system from the influence of societal norms. Conversely, the
advancement of knowledge is not stifled, and in fact may often be enhanced, when basic research is performed with an eye toward potential
applications.

Researchers motivated by curiosity about nature have produced a great
abundance of startling, unexpected, and marvelous discoveries over the
past fifty years (many of which have proven to have unanticipated practical applications). Such discoveries are the products of the brains, curiosity, creativity, and luck of scientists. But they are also the products of
cultural, political, economic, technological, and even, it appears, sexual
context. These are very real constraints on research. In portraying a world
in which basic research exists apart from society, the myth of unfettered
research may therefore be viewed at its root not as a justification for
protecting scientists from the whims of politicians and voters—indeed,
the federal R&D system is a product of political action—but rather as a
rationale for preserving the existing power structure and priorities of the
system. The claim that social context is irrelevant to basic research merely
strengthens the prevailing context. Uncritical acceptance of the myth may

*An analogous argument can be advanced about underrepresented ethnic and racial minor-
ities, of course. The whiteness of the R&D system is as distinctive a characteristic as its
maleness.

therefore have the paradoxical effect of discouraging the creation of new knowledge in neglected areas of science, concentrating new expenditures on areas that are already generously supported, stifling democratic discourse over research priorities, and insulating the basic research system from social and political accountability.

The Myth of Accountability

Without maize, the giant Mayan or Aztec pyramids, the cyclopean walls of Cuzco or the wonders of Machu Pichu would have been impossible. They were achieved because maize virtually produces itself.

The problem then is that on one hand we have a series of striking achievements, on the other, human misery. As usual we must ask: who is to blame? Man of course. But maize as well.—Fernand Braudel, *The Structures of Everyday Life*

THE MYTHS OF INEVITABLE BENEFIT and unfettered research do indeed imply that social accountability is inherent in the research process. If more research means more societal well-being, yet no particular line of research is more likely than any other to contribute to this well-being, then accountability is a strictly technical matter and the delivery of benefit to society is simply a question of how much knowledge the research system produces. A scientist's responsibility to society is therefore clear: to conduct work of high intellectual integrity. The broader responsibility of the research system as a whole is equally straightforward: to ensure that the products of science conform to the highest possible scientific standards. Society must be satisfied with the science that it supports, so long as that science meets the criteria of integrity established by the research community. At the same time, because only scientists are qualified to apply these criteria, any effort by politicians or other outsiders to assess the effectiveness of research is pointless. To the extent that such efforts may impede the production of new knowledge they are also counterproductive.

Nevertheless, in an age when science and technology have become integral components of democratic society, it is inevitable that politicians and voters will seek to influence the shape and direction of an R&D system that is supported by the government. The seeds of political conflict are thus intrinsic in the federally funded research system. The battleground for this conflict is the issue of accountability, and many scientific leaders believe that the battle has begun.

This story has several versions. One is that a scientifically illiterate public is being unwittingly exploited by the religious right or the environmental left to bolster political agendas that fly in the face of the teachings of science. Another is that the press is distorting and misrepresenting the character and attributes of the research system to a credulous and scandal-mongering public and thereby turning the tide of public opinion against science and scientists. A third is that humanists and social scientists of the academic world, operating in the spiritual void of postmodernism, are perversely intent on dragging the natural scientists into that void as well. The particular tragedy of these trends, argue leading voices of the scientific community, is that today, more than ever in the past, the world needs the knowledge, the know-how, the objective perspective of science in order to confront the global challenges facing an increasingly global society. The questioning of science—its value, its appropriate role in modern culture—is commonly portrayed not merely as a reflection of ignorance but also as a threat to humanity's prospects for the future. One extended quotation will serve to summarize the general position:

> [Science] must be valued by society because science claims to be the very embodiment of the classical values, starting with truth, goodness and beauty. . . . Thus, science has been widely praised as a central truth-seeking and enlightening process in modern culture. . . . Science also has been thought to embody the ethos of practical goodness in this imperfect world. . . . Finally, the discovery of beauty in the structure, coherence, simplicity and rationality of the world has long been held up as the ultimate, thrilling reward for the innovator as well as for the student. . . .
>
> Most of the optimistic description I have just given was widely taken for granted during recent decades, embodied for example in

the famous Vannevar Bush report of nearly 50 years ago, the main driving force of science policy thereafter. And much of it can still be demonstrated convincingly. . . .

[But the] general balance that had been achieved during the past few decades is changing precipitously . . . and with it a whole range of relations between science and society, hence between science and values. . . . For what has entered into the equation commanding the up and down motion of the lever of sentiments is an agent, a weight unlike any in the whole history of the rise and fall of the perceived value of science itself. . . .

What is the new agent that has entered? . . . The first element is an assertion, mounting louder and louder over the past few years in books and hundreds of articles. . . . I refer of course to the widespread assertion that the pursuit of science, to a previously completely unrealized degree, requires us not merely to reassess constantly the safeguards on its ethical practices and uses—of that, there is a long tradition in the scientific community . . . but that the pursuit of science is, and has been all along, since at least the days of Hipparchus and Ptolemy, thoroughly corrupt and crooked; and that consequently severe measures must be applied to the practice of science from the outside.[1]

The author, a distinguished historian of science and a physicist, sees an "emerging sentiment" that spells the "delegitimation of modern science as a valid intellectual force."[2] The consequence of this "delegitimation" is portrayed as a destructive transition in the nature of scientific accountability, from internal mechanisms rooted in "safeguards on [science's] ethical practices and uses . . . in the scientific community" to "severe measures . . . applied to the practice of science from the outside."[3]

Other, more fragmentary examples may help to flesh out the terms of engagement as perceived by the research community:

"[More] Americans know their astrological sign than understand that astrology is superstition. And only a tiny slice of the population grasps the scientific complexities of tomorrow's issues rather than the simplistic arguments hawked, for example, by those who brought us the 'dangers' of Alar in apples, the supposed imminence—20 years ago—of a new ice age, and global famine by 1990."[4]

Another perspective refers to "a populist anti-intellectual impulse in the United States that is being cynically encouraged by the radical right. That impulse manifests itself in diverse forms, such as attempts to get creationism taught as science, a general assault on colleges and universities as being too 'liberal,' and attacks on the credibility of environmental scientists who are attempting to inform humanity of the perils it now faces."[5]

In addition, Congress and the press are excoriated for their investigations into allegations of scientific misconduct and misuse of federal research funds. "I have a very real concern that American science can easily become the victim of this kind of government inquiry," said Nobel Prize–winning biologist David Baltimore, who was one subject of a widely publicized congressional investigation. "Uninformed or malinformed outsiders cannot effectively review the progress of scientific activity."[6]

Finally, a former president of the American Association for the Advancement of Science sees "a cause for great concern over the role of science in a democracy in which the general population has not enough understanding of science itself; does not entirely trust 'science experts' and does not want to; and is left with no way to distinguish between the competing claims of apparent experts on both sides of any question."[7]

The consequences of such dynamics are thought to be catastrophic: "Some non-scientists think that Science is pernicious, that it is responsible for the pollution, contamination and despoliation of our air, our water and our earth. Tried *in absentia* and found guilty, the scientific enterprise should be terminated."[8]

These and many similar perspectives add up to a portrayal of a society that is increasingly anti-intellectual, scientifically illiterate, wrongheaded, and easily manipulated. These trends are said to be exacerbated by the press, the government, and academic disciplines outside of the natural sciences that ought instead to be embracing science as the key to a better future.

But this portrayal is undermined by an examination of public attitudes about science, as revealed in surveys taken over the past several decades. In 1957, 94 percent of Americans agreed with the statement: "Science and technology are making our lives healthier, easier, and more comfortable." Although this level of support had declined somewhat by the late 1970s, it

stabilized at about 85 percent during the 1980s. Of the fifteen industrialized countries in which surveys were administered in 1990, not one
yielded a significantly higher level of public confidence in science than
the United States (83 percent). Lowest among the surveyed nations was
Japan—the world's technology titan—where only 54 percent of the public
agreed with the assertion that science and technology improved the quality of life.[9]

Surveys do confirm that rates of scientific literacy among Americans
are abysmal, but this is no recent phenomenon. Literacy rates have been
more or less constant since their systematic measurement began in 1957.[10]
According to one 1988 survey, about 6 percent of the public can be
adjudged scientifically literate, although about 50 percent of those surveyed showed an "understanding of the impact of science and technology."[11] In overall measures of science literacy, Americans rank about in
the middle relative to citizens of other industrialized nations, although of
fourteen surveyed nations, only in the United States do a majority of
people (60 percent) agree that astrology is "not at all scientific."[12] In
comparison to other nations, a much smaller minority of Americans (14
percent, versus 37 percent for Europeans) believe that "it is *not* important . . . to know about science in . . . daily life."[13]

Despite high levels of scientific illiteracy, Americans strongly support
the federal commitment to research. In 1992, 76 percent of the public
agreed that basic research, "[even] if it brings no immediate benefits, . . .
should be supported by the Federal government." Seventy-three percent of Americans—a far greater proportion than in Europe or Japan—
believe that the "benefits of science are greater than any harmful effects."
Seventy-nine percent of Americans believe that "most scientists want to
work on things that will make life better for the average person."[14] Even
the controversial area of biotechnology seems to enjoy strong support
among U.S. citizens.[15]

The public seems to share with scientists a general contempt for the
press. Between 1973 and 1993, the percentage of Americans who had "a
great deal of confidence" in the "people running" the press declined from
23 percent to 11 percent. Public confidence in the leadership of scientific
institutions has hovered at about 40 percent over that same time interval,
second only to confidence in the leadership of the medical community.[16]

These results conflict in several ways with the views commonly expressed by the scientific community. First, the claim that there is a major antiscientific sentiment sweeping the United States is wrong; the level of general public approbation for science and scientists is still overwhelmingly high. Second, Americans seem acutely aware of the importance of science and scientific knowledge, and they seem highly supportive of scientific research, including basic research. Third, rates of scientific literacy have been low for decades, so they cannot be the cause of perceived changes in public attitudes toward science.

While leaders and spokespersons of the scientific community believe that the research system is under assault by the laity, there remains in fact a remarkable degree of public support for science. What, then, is the source of the apparently widespread conviction among scientists that their once happy marriage with the public, with government, and with the press is now on the rocks?

The pervasiveness of science in people's lives is manifest in political institutions, in the media, in business, education, the home, even in religion. Every phase of human activity is affected: work, recreation, travel, communication; eating, drinking, procreating, dying. Public debate these days, regardless of the issue at hand, will invariably include some scientific element because science bears on a comprehensive range of public concerns in modern society. Scientists, meanwhile, are fighting to justify and preserve a relationship with society based on isolation from the realities they have helped to create. Science has entered the realm of participatory government and everyday life and will not easily extricate itself so long as it makes a claim on public funds and promises to deliver public benefit. This, perhaps, is the price of success—at least in a democracy.

Viewed from a policy perspective, science is joining other major institutions, such as the military and business, whose leaders would very much like to be ignored by the political system but whose activities must be monitored and debated and regulated and held accountable through the political process exactly because they have such a significant impact on the whole of society.

Leaders of the research community have almost universally failed to grasp the nature and inevitability of the political change that is overtaking their enterprise. They continue to view the question of accountability as a housekeeping problem. Scientists have focused much concern on several highly publicized government investigations into scientific misconduct and misuse of public funds. Two affairs garnered particular attention: the accusation that a publication coauthored by biologist David Baltimore was based in part on falsified data, and allegations that Stanford University, the leading academic recipient of federal research funds, engaged in questionable accounting practices such as inflation and misallocation of overhead costs charged to the federal government. Some scientific leaders saw these cases as examples of unjustified government meddling into the conduct of science, arguing that the government's involvement was motivated in part by ignorance and in part by politics and that it resulted in irreparable damage to the research system. Others saw these incidents as reasonable exercises of government jurisdiction that demonstrated the need for the recipients of federal research funding (both individuals and institutions) to be more sensitive to the appearance and reality of impropriety.[17]

The National Academy of Sciences issued a report on scientific misconduct that embodied a rather bland middle ground:

[Statistics] indicate that the reported incidence of misconduct in science is low. . . .

However, any misconduct comes at a high price both for scientists and for the public. Cases of misconduct in science . . . breach the trust that allows scientists to build on others' work, as well as eroding the trust that allows policymakers and others to make decisions based on scientific evidence and judgment, especially in instances when definitive studies are not available. The inability or refusal of research institutions to address misconduct-in-science cases can undermine both the integrity of the research process and self-governance by the research community.[18]

However, a member of the panel that produced the Academy report viewed the problem in a more political context:

Today science is perceived by some as yet another interest group whose claims to public funds must be severely scrutinized especially in light of questions regarding the funding and conduct of university research.

These questions cannot be dismissed lightly. Science is essential to the solution of many of the world's problems, and it is vital that the public's esteem for and trust in science is maintained. The few, highly public cases of scientific misconduct have brought into question the university's ability to manage the research done on its campus and the scientific community's ability to warrant the public trust. While there is no evidence that the scientific knowledge base has been seriously affected by these cases, the universities and the scientific community have been damaged in the eyes of the public and the Congress—not so much because they occurred, but because they were not well handled.[19]

Because the internal integrity of the research system was called into question, many researchers apparently perceived the Baltimore and Stanford cases to be principal symbols of the growing stress between science and society. But these cases are no more than marginal consequences of the evolutionary but fundamental changes overtaking the science-society relationship. The scientific worldview does not encompass mechanisms by which the research system can respond to such changes. According to this worldview, scientific accountability *is* societal accountability: the checks and balances inside the research system—peer review, open scientific debate, reproducibility of experimental data, and testing of alternative hypotheses—guarantee the integrity of the system as a component of society.

This is the myth of accountability: that internal integrity equals external responsibility; that the ethical obligations of the research system begin and end with the delivery of a scientific product that is quality controlled and intellectually sound. Consider the following analysis:

> [We] would really be quite incredulous if one day soon Congress abolished its sewer-money financing of campaigns. . . . We would think it a premature April 1st joke if we were told that the American tobacco industry or the National Rifle Association were planning to [hold] an open meeting . . . on the ethics and values of their enterprises. Instead, [critics] are more likely to focus on a group that, precisely because of its avowed incorporation of its own honor code, and because of its preference for organizing itself only poorly for

public purposes, is singularly vulnerable to having guilt feelings induced in them: I speak of course of our research scientists. Among these, the rate of publication of provable fraudulent or falsified data has by one estimate been down at the astonishingly low level of around 0.0002 percent, owing chiefly to the internal mechanisms of validation.[20]

In many respects, however, a debate over the scientific validity of an obscure aspect of bioinorganic chemistry or paleoseismology is similar to an American tobacco industry discussion about the best strategy for marketing cigarettes or a National Rifle Association controversy over the safest and most effective way to kill deer.[21] In each case, participants in these debates share a frame of reference that allows them to communicate with each other but largely excludes the outside world. Scientific debate is precisely analogous to debate that occurs between members of other groups whose shared assumptions, goals, and language isolate them in some way from the rest of society. The "honor code" to which scientists are expected to adhere is about the conduct of science, not about the "ethics and values" of science as a component of society.

Indeed, one effective function of the myth of accountability is that it necessarily excludes outsiders from the process of judging science or scientists. Traditional accountability in science rests on the capacity of experts to judge the methodological and substantive validity of each other's work. In practice, because research in most areas of science has become tremendously specialized and arcane, only a relatively small number of researchers in any particular subdiscipline is competent to evaluate work within that subdiscipline; no one else has sufficient exper-tise to render judgment. Thus, when scientists call for more scientific literacy among the voting public, can they possibly intend that this liter-acy should lead to greater public inclusion in the internal accountability process? They probably expect the opposite: that the public—if it became fully aware of how science works, of how scientists live by an "honor code" and impose internal accountability through peer review and other mechanisms—would be fully content to provide unconditional, unques-tioning support for the scientific community. Of course, researchers might be surprised by the attitudes of a more scientifically literate public. For example, survey results from Europe show that those nations with the highest rates of scientific literacy also display the highest degree of skepticism about the benefits of science and the judgment of scientists.[22] There is no reason to believe that more science literacy will lead to less questioning and debating of the appropriate role of science in society.

From a political standpoint, the accountability problem that the scientific community faces is rather simple. Public support for science is justified in the context of public benefit. Science is portrayed as the key to solving the great problems that face society today: economic stagnation, environmental degradation, resource depletion, epidemics, overpopulation. These are promises, publicly made, often repeated, long remembered, and, most importantly, not easily delivered.

Government funding for research and development implies the existence of a social contract: In return for public funding of much of the nation's research, science is expected to deliver benefits to the public. But how can science and scientists be held accountable for fulfilling their end of the bargain? The contract is based on promises made to society on behalf of science, yet it includes no mechanisms to measure how successfully these promises are being met. The system of internal scientific accountability is ill suited for this task; it is not designed to monitor the effective contribution of science to societal goals. There is, in other words, an accountability gap, a loophole in the contract which practically guarantees that the scientific community, in seeking to justify public expenditures on research, will overstate its claims of public benefit.[23]

Former National Academy of Sciences president Frank Press confronted the problem in an interview with the magazine *Physics Today*:

Q. [It] sometimes happens that a project or program is oversold—that is, a great deal of hype is used to convince politicians, the press and the public that it's worth doing. Aren't there dangers in making excessive promises and raising expectations for what science can do for society in the way of applications or products?

A. It's unbecoming. It's even intellectually dishonest for scientists to make such statements purely for the purpose of getting funds. And to the extent that this happens, it detracts from science as a credible intellectual endeavor. . . .

Q. Are you saying there is something inherently unethical about hype in science? . . .

A. You have to be careful that you don't get into the issue of the proper behavior of scientists as they perform their work honestly, within the culture and the ethic of scientific method. . . . I don't consider hype to be unethical in the sense of scientific dishonesty.

Q. Is it any different from false marketing of products?

A. Look, I don't like it and you don't like it, all right? What I'm saying is that it's not scientific misbehavior, and I wish it didn't happen.[24]

This is a fascinating exchange for what it reveals about the scientific view of ethics and accountability. Although Press acknowledged the existence of an accountability gap and criticized the rhetoric that led to it, he was extraordinarily careful to avoid labeling such behavior as unethical because in doing so he would have expanded the concept of accountability to embrace a broader view of social responsibility. Furthermore, he portrayed "hype" in science as a recent trend arising from increased competition for federal research funds: "The only way to understand this phenomenon is that scientists are not themselves because of the crisis in funding and their race for grants."[25]

There *is* another way to understand this phenomenon, however. Perhaps what was once accepted by the public as credible is now received with a more critical eye. The words used by Vannevar Bush in 1945 to explain the societal benefits of science may not be so different from those heard today, but the social and political context in which those words were delivered has changed profoundly. The promise of science illuminated by the glow of America's World War II victory may now be heard as the hype of science in postindustrial society. As people become more aware of the impact of science on their lives, it is inevitable that they will compare their perceptions of the state of society—and of changes in collective and individual quality of life—with the rhetoric used to promote science. In doing so, they will recognize that advances in science do not equal advances in human affairs. In the face of the invariant claims of the scientific community—more science means more public benefit—increased public scrutiny and skepticism may be inevitable.

In responding to such changes, the scientific community has hastened to assure the public that its integrity is still sound, that the quality of the scientific product is still of the highest order, and that the David Baltimore and Stanford University incidents are mere aberrations. But such reassurances verge on the immaterial, like a sailor rushing to tell his captain that their ship, foundering upon a reef, has plenty of fuel left in its tanks. Broader concepts of scientific accountability are not being consciously imposed by Machiavellian legislators in response to a few instances of bad behavior in the laboratory. They are arising spontaneously from a society that finds itself rife with uncertainty and insecurity at the same time as its knowledge of the natural world—and its ability to manipulate that world—grows ever more potent.

Research directions chosen by the scientific community often do diverge from the knowledge needs of society. The belief that high-quality research is sufficient to ensure the creation of social benefit has not, as we have seen and will continue to see, entirely withstood the test of time. In the case of American manufacturing, for example, many U.S. industrial leaders assert that while the American research system excels at exploring the frontiers of knowledge, it nevertheless neglects lines of research necessary to permit the application of existing knowledge to manufacturing problems. One leading research manager suggests: "[America's] problem is dominated by inadequate attention to the use of what we know, and inadequate use of the skills and the technical competence of those who have the knowledge of science and technology. . . . This does not mean that we need a pause in knowledge generation . . . but only that . . . the more serious problem lies with the use of knowledge."[26]

An analogous argument has been made for the biomedical system. A report of the World Bank suggests that the basic biomedical research agenda of the industrialized world has an indirect, adverse affect on health care in developing nations because it focuses on problems of marginal social import—even if of compelling scientific interest—and thus diverts attention and resources from more useful but less scientifically attractive work in epidemiology, preventive medicine, and other areas of applied health sciences.[27] Recent criticism of progress in AIDS research has underscored a similar theme, suggesting that scientists have been attracted to research on highly esoteric aspects of the disease while neglecting less glamorous research that would focus on applying existing knowledge to AIDS prevention.[28]

One particularly well-documented example of the divergence between research trends and social needs was the National Acid Precipitation

Assessment Program (NAPAP). This research effort, which ran from 1981 to 1990, was created to help the government assess the environmental threat of acid rain and to develop an appropriate legislative and regulatory mitigation strategy. NAPAP failed to achieve its goals in spite of the scientific excellence of the research performed under its auspices. This failure arose from the inclination of researchers to view accountability purely in terms of internal, or scientific, criteria. So long as the science was performed in adherence to the internal standards of the research community, scientific responsibility to society was seen as fulfilled. Thus, the principal motivation of the researchers who administered and carried out the program was to generate interesting, high-quality science, even if such science was of no particular use to society as a whole: "NAPAP scientists built the biggest models, conducted the most extensive surveys. . . . NAPAP research elucidated some of the intricacies of the atmospheric mixing and transport of various pollutants. . . . But in the end, the program never got around to the nuts and bolts questions about the costs and benefits of control strategies."[29] Why did this occur? As one study explained, "[No] serious effort was made to define policy-related research priorities and to then shape an appropriate set of projects and timetables to answer these questions in a meaningful way. Instead, independent government agencies, driven by different missions, motivations, and expertise, largely pushed their own scientific research agendas."[30] One cause of this problem was that "research was managed and reviewed largely by researchers interested in pursuing scientific research, not applied, policy-related studies."[31] Similar problems have been predicted for the U.S. Global Change Research Program, whose much more complex and expensive research agenda, as the next chapter discusses, is driven primarily by the curiosity of scientists rather than the needs of policy makers.

In general it can be expected that, when given the option, the research community will choose scientific criteria over societal ones in determining the overall direction of its work. The myth of accountability can be seen as a rationale for insulating the prerogatives of the scientific community from the needs of society in the name of "scientific excellence." This argument seems particularly clear for basic research, whose advocates turn the concept of accountability on its head by arguing that the best way to ensure the resolution of concrete societal problems is to support research on the basis of scientific merit alone. From this perspective, society actually becomes accountable to the research system: failure to provide "enough" funding for basic research will translate into a failure to solve societal problems—even though no one can predict which problems will be solved—while politically motivated efforts to influence the direction of the research system will necessarily reduce the capacity of science to contribute to society.

Thus, when society demands a level of accountability from the research community that goes beyond the internal quality control process that already exists, leading scientific voices may portray these demands not merely as antiscientific but also as a violation of societal self-interest. A typical pattern in these types of controversies is the belittling of public opinion and the insistence that all controversy is engineered by extremists or charlatans who manipulate ignorant politicians into taking action that is explicitly contrary to the greater public good.

The case of biotechnology is illustrative. In the early 1970s, a group of prestigious molecular biologists took the highly unusual step of calling for a temporary, voluntary moratorium on research that made use of new recombinant DNA techniques because of concern that "some of these artificial recombinant DNA molecules could prove biologically hazardous."[32] This action demonstrated that some professionals in the research community perceived a direct link between their activities in the laboratory and their ethical responsibilities to society. As these concerns were gradually allayed by continued study, recombinant DNA research resumed, and it now forms one of the crucial techniques of the biotechnology revolution and the mapping of the human genome.

Although members of the scientific community were undeniably concerned about the unforeseen consequences of recombinant DNA research, scientists often greet public concerns about such research with disdain. A mere four years after signing the letter advocating a research moratorium, biologist David Baltimore was attacking public opposition to the same research: "We were addressing a limited problem, whether there could be a recognizable hazard in the performance of certain experiments. That limited question opened a floodgate; other questions came pouring out and are still coming. . . . The mayor of Cambridge, Massachusetts, raised the specter of Frankenstein monsters emerging from MIT and Harvard laboratories, and speculations about the possibility of inadvertent development of a destructive organism like the fictitious Andromeda Strain have been much in the news."[33] Some scientists came to regret their flirtation with public accountability: "There is a need to overcome the lingering public anxiety created when scientists declared a moratorium on recombinant DNA research. . . . The public understood the moratorium to mean that biotechnology might possess unique hazards and it is now very difficult to undo the harm done by the public impression that biotechnologists are about to release the 'Andromeda Strain.' "[34] Thus is the moratorium on recombinant DNA research now portrayed as a cause of "harm" because a debate that the research community wanted to keep inside the walls of the laboratory leaked out into the real world. While scientists could be legitimately and rationally concerned about "new kinds of hybrid plasmids or viruses, with biological activity of unpredictable nature,"[35] public anxiety over the "Andromeda Strain" was nothing more than the hysteria of the ignorant.

On its face, the question here is one of specialized knowledge. When scientists ridicule public opinion, they implicitly argue that wise decisions about the potential societal impacts of research can only be made by experts. But if, as researchers say, the results of their work are intrinsically unpredictable, then experts are no more qualified to assess the potential for negative societal consequences than any other reasonably informed person. If, on the other hand, such consequences can be generally foreseen, then the public has a right and an obligation to be involved in deciding what sorts of actions should be considered; technical knowledge can guide but not dictate these decisions. Either way, the question of technical expertise is secondary. Of primary importance are the underlying social and moral dilemmas that drive public concern. As one commentator suggests: "People who call themselves opponents of biotechnology . . . are often not so opposed to the technologies themselves as they are to the social, industrial, and political systems that may be created around them."[36] But it is these very issues that the research community is loathe to acknowledge, for in doing so they would open the floodgate of public involvement.

The reluctance of the research community to engage the social implications of its work was particularly conspicuous in the controversy over fetal tissue research, which officially began in March 1988 when the U.S. Department of Health and Human Services (HHS) banned federally funded investigations into the treatment of human adults with transplanted tissue from induced aborted fetuses.[37] This moratorium stayed in effect until 1993, when President Clinton, as one of his first acts in office, reversed the policy.

The fetal tissue controversy has little to do with science and everything to do with the ethical and political debate over abortion. Presidents Reagan and Bush, who counted on the political support of antiabortion groups, supported the ban; President Clinton, a prochoice advocate, opposed it. Underlying these political considerations is perhaps the most protracted, intractable, and divisive moral debate to occur in the United States since the Vietnam War. If one believes that abortion is murder, then any use of the tissue from induced abortions is morally unacceptable, however laudable the goal. If one believes that abortion is a personal decision that lies in the hands of individual women, then fetal tissue research is unobjectionable.

Not everyone in the research community managed to catch on to this undercurrent. Following the initial moratorium in 1988, the National Institutes of Health appointed an advisory panel of experts to study the problem. Not surprisingly, the panel concluded that fetal tissue research was "ethically justified."[38] Louis Sullivan, Secretary of HHS, ignored the panel and extended the ban. Scientists were outraged. One researcher said, "It's like the Middle Ages. We are very disappointed. This ban interferes with research, with new knowledge that is going to save the lives of fetuses, babies and adults as well."[39] Another said, "I know of no precedents of repression of Federal science like this one."[40] A third scientist argued explicitly that the decision was one that should be made by the scientific community: "It's a matter of whether the community wants this kind of research to be done. If they don't want it done I don't think we should do it. But I don't think this guy [Sullivan], with all due respect, speaks for the community as a whole."[41]

Scientists were unwilling to accept that the roots of the fetal tissue debate were not in their bailiwick. Those scientists who strenuously objected to the moratorium were presumably supporters of the prochoice position, but this connection never became a visible part of the public debate. Outrage was focused on the issue of interference with the research process, not interference with a woman's right to choose abortion. As articulated by the science community, the fetal tissue controversy was a battle over scientific freedom, not a struggle to define ethical norms. One researcher, in explaining the potential application of fetal tissue research to the treatment of infertility, wrote: "It is entirely proper in a democratic country that ethical questions involving science are resolved by society as a whole rather than by scientists alone. But, as one who confronts the heartbreak of infertility every day, I wish more Americans would recognize what we could accomplish by pursuing this research."[42] Careful reading of this statement suggests underlying sentiments that are precisely contrary to its apparent meaning. Use of the word "but" reveals that the author does not entirely believe what he says about the resolution of ethical questions in a democracy. The debate over whether the research should be carried out is given precedence over the underlying ethical debate over abortion. The author uses his own subjective experience ("as one who confronts the heartbreak of infertility every day") to create a second-order moral position that appears to address the societal issue but in fact explicitly avoids it.

The language used by scientists in favor of fetal tissue research was built upon, but failed to acknowledge, a nonscientific moral position— that a woman does have a right to choose to have an abortion. This assumption was not a visible part of the debate for a very good reason: by acknowledging that the scientific position was built upon an ethical and therefore subjective foundation, the research community would be acknowledging a level of accountability to society that it finds congenitally repugnant. Thus, the rhetorical structure of the scientific position vaulted over the fundamental moral debate, began with the assumption that the debate was about research, and built a subsidiary moral position from there: If you suppress research, then you suppress the benefits of research as well. Implicit in this argument is the view that moral standards established by the R&D community can be insulated from those of society at large and that the research enterprise should be accountable only to itself and not inhibited by the moral dilemmas faced by the outside world.

On the whole, public attitudes about science appear to be more balanced and nuanced than the attitudes expressed in the scientific community about society. Science enjoys the high regard of most Americans; broad public support for basic research suggests that the average citizen is adequately sensitized to the importance of intellectual freedom and scientific curiosity in the laboratory. The public believes that its future well-being will depend in part on scientific advance. But the infiltration of science into every corner of modern culture ensures that both science and scientists will evermore be ensnared in a broad range of political controversy and, therefore, that they will be subjected to the same sorts of political indignities suffered by every other organized component of democratic society. One of these indignities will be the demand that the research enterprise—its conduct, its products, its consequences, and its rhetoric—be held accountable to priorities and norms established by society, as vague and changeable as these norms undoubtedly are.

The scientific community understandably resists such trends, and its leaders maintain that any effort to exact a measure of accountability to public standards is ill conceived and rooted in ignorance or politics. "All such rhetoric can do is hurt even more the efforts to support good science in this country," said one observer.[43] Efforts by the National Institutes of Health and the National Science Foundation to develop new mechanisms of accountability to society through fuller articulation of strategic research goals have failed to win the support of the research community.[44] Scientific leaders maintain that the contribution of science to society cannot be improved by any changes in the way that science is carried out; that the system, while not perfect, is about as good as it can be.[45] As one researcher explained: "The things politicians are worried about are not so, and the things they want fixed are not broken."[46] The inescapable extension of such arguments is that it is appropriate for science to have a profound and irreversible impact on the course of society but inappropriate for society to exercise jurisdiction over how science goes about creating this impact.

If scientific accountability to society goes beyond the mere assurance of integrity inside the laboratory, then it raises fundamental questions about the organization and administration of the research system as a whole. If the scientific community bills itself as a major contributor to human welfare—as it does—then this billing must be tested against the course of societal evolution. When leaders of the research community perceive themselves to be under assault by the forces of scientific illiteracy and political extremism, they might be encouraged to explore alternative hypotheses: to look inward, at the rhetoric and conduct of the research system. After all, it is the indulgence of a 94-percent-scientifically-illiterate society that has allowed that system to grow and flourish for half a century. Of course political extremism and anti-intellectualism are dangers to science, just as they are dangers to democracy. But the contempt that many leaders of the scientific community express for the public and the nonscientific institutions of Western society may suggest another type of danger.

When scientists portray all critiques of science as the products of uninformed, irrational, or politically motivated thinking, then they portray themselves as beyond social accountability. Indeed, this is the essence of the defensive posture adopted by many vocal scientists today: "In the days of my youth, science, scientists and the quest to understand nature were unquestionably good. Now, however, according to a clutch of science critics—journalists, philosophers, politicians, and simple science bashers—science is no longer good, nor are scientists."[47] If one believes oneself to be "unquestionably good," then questions must be bad. If questions are bad, then there is no basis for democratic dialogue between scientists and the public. But why should a nonscientist ignore the fact that science and technology, while promoted as the cure for a thousand societal ailments, have at least in some part been a contributor to many of these ailments as well? Why should a nonscientist fail to wonder at the social chaos and insecurity of a culture that is ever more tightly bound by technological marvels of communication and transportation? In the face of such questions, rooted in apparent contradiction, it may be both politically and intellectually unreasonable to demand that the public should accept science's self-proclaimed role as humanity's savior without seeking to impose some measure of influence over the path to this salvation.

5

The Myth of Authoritativeness

It is easy to laugh at anyone for being wise after the event, but it is almost as useless to be wise before it.—Italo Svevo, *Confessions of Zeno*

BUT WHERE IS THE COMMON GROUND between the orderly search for scientific truth and the chaotic forums of popular governance? As science and technology grow ever more pervasive in daily life, shouldn't this common ground be sought in a more scientific approach to democracy rather than a more democratic approach to science? A 1990 editorial in *Science* illustrates the problem by means of an imaginary dialogue between a confused but well-meaning ingenue named "Science" and a sort of philistine Everyman called "Dr. Noitall":

SCIENCE. [It] is our job to tell people when 2 + 2 = 4.

DR. NOITALL. That's exactly where your views are wrong. A recent poll shows that 50% of the people think 2 + 2 = 5, and almost every network agrees with them. Those people have rights, they believe sincerely that 2 + 2 = 5, and you take no account of their wishes and desires. Simply imposing 2 + 2 = 4 on them is not democracy.

SCIENCE. But there is really no serious disagreement on the question. . . . We can't take seriously people who make emotional rather than scientific arguments.

DR. NOITALL. That reflects a condescending attitude toward those who did not have the privilege of having an advanced education. Prominent political groups have already supported enactment of legislation, even if it is scientifically inaccurate, as long as the public wants it. . . . There are two truths in this world: one of the laboratory, and the other of the

71

media. What people perceive as the truth is truer in a democracy than some grubby little experiment in a laboratory notebook. A stubborn insistence on the facts instead of people's perception of the facts makes you look heartless and disdainful.[1]

This didactic exchange, appearing in America's leading scientific journal, encapsulates a perspective about the relation between science and politics that seems to be widely held within the scientific community. If politics is "the arena of the irrational,"[2] then science is the only rational player in the game and scientific truth must be a guiding parameter for wise political decision making. According to this perspective, science could resolve a multitude of the thorny political problems that confront people and nations if only humanity would rely on "facts" instead of "perception of the facts." Scientists could introduce the thread of objectivity into the crazy quilt of political interests that so often smothers society's earnest efforts to come to grips with its problems: "In an ideal world, politicians would turn to scientists every day for advice on the choices facing government. These might concern the felling of rain forests in Brazil and Africa, and all that this implies for the greenhouse effect; the side effects of a new drug; the long-term action of pesticides; or how best to exploit oil from the North Sea or the Gulf of Mexico."[3]

In many ways, this "ideal world" has actually come to pass. In recent decades the United States government has created a plethora of mechanisms through which it can solicit expert advice on scientific and technological questions. Scientists hold advisory positions at every level of government, in the executive agencies, and in the Congress. Nongovernmental organizations such as the National Research Council issue dozens of in-depth studies each year aimed at giving politicians and bureaucrats the information they need to make better decisions on controversial issues having a technical component. Thousands of scientists voluntarily offer their expertise as members of various governmental advisory committees and as expert witnesses testifying before Congress. Finally, the government spends literally billions of dollars supporting research intended to provide scientific guidance to the political decision-making process. Leading the way in 1995, for example, was the $1.8 billion U.S. Global Change Research Program, but there are countless smaller programs as well, in areas ranging from construction standards to public health.

Still, the government often seems unwilling or unable to make wise use of all this advice and data. Government scientists have been fired for publicly supporting scientific hypotheses that fly in the face of prevailing political dogma; others are forced to publicly support political positions that conflict with their own scientific convictions. Policy decisions in areas such as environmental cleanup and biomedical research are often criticized by scientists as excessively political and inconsistent with scientific knowledge. Overall, it would be very difficult to point to any major political controversy that has been resolved on the weight of authoritative scientific data. Political concerns and public opinion do seem to outweigh the input of scientific experts.[4] The culprit, asserts one Nobel Prize winner, is "our political processes, which time and time again have ignored the prescriptions emerging from scientific research."[5] And the problem may only get worse. An increasingly technological society finds itself facing an increasingly complex array of scientific and technological challenges in areas ranging from energy policy to health care policy, high technology industrial development to computer network regulation, toxic waste disposal to global climate change mitigation and response. The satisfactory resolution of such challenges, with their vexing mix of the political and the technical, is commonly said to depend on the ability of government to make rational policy decisions that subjugate political expedience to scientific truth.

There is no dearth of ideas about how to create a political environment that is more open and responsive to scientific advice. Such ideas commonly emphasize that more scientific information needs to find its way into the political arena. "The quality of congressional decisions on [scientific and technical] issues depends on the quality and usefulness of information and analysis made available to Congress by scientists, engineers, and others," said one report, which included recommendations such as: "Congress [should] improve its approaches to obtaining S&T [science and technology] analysis and advice from the scientific community," and "more scientists, engineers, and others [should] become actively involved in science and technology policy activities."[6]

What is the thinking behind such unobjectionable goals? What is it about scientific and technological advice that is supposed to cut through the self-interest and rhetoric and short-term thinking associated with governing? That is, how is science supposed to save us from politics? Apparently these sorts of questions are considered so obvious that they do not merit any discussion because they are almost never addressed in the multitude of calls for more and better science advice. Science seeks truth; science pursues objective knowledge; science is not influenced by political interests or short-term considerations or emotion. Science, in other words, is supposed to be everything that politics is not. Thus, it is said, if our political institutions fail to grapple effectively with a difficult technical issue, then the problem must in some manner be traced either to inadequate communication between scientists and politicians or to the irresponsible actions of politicians who ignore "the prescriptions emerging from scientific research."

Fundamental to virtually all suggestions for improving the quality and quantity of governmental science advice is the belief that such advice can have a direct positive impact on policy making—that the authoritative voice of science would be translated into wise policy responses by the government if only politicians took actions consistent with the advice and information provided by scientists. This is the myth of authoritativeness, and it rests on two assumptions: first, that the intrinsic value of scientific information must, of its own weight, improve the ability of governments to make effective policy decisions and, second, that there is such a thing as authoritative scientific information which, once recognized by politicians, can be applied to the major policy challenges facing governments today.

The policy model goes something like this: (1) The government is faced with a difficult political problem that contains a significant technical element. Examples might include the need to achieve energy independence or to dispose of radioactive waste or to control acid rain. (2) Congress and the appropriate federal agencies call on the technical experts inside and outside the government for advice. (3) The experts transmit information to the government. If more information is needed, then the government supports additional research. (4) The appropriate laws or decisions are made to resolve the problems.

Steps 1–3 seem to work with a reasonable degree of success. Questions do get asked, expert advice is solicited and heard, additional research is conducted. Yet the model seems to break down at step 4, as the information provided by scientists often fails to translate into decisive political action. One explanation for this failure was offered by Dr. Noitall: "What people perceive as the truth is truer in a democracy than some grubby little experiment in a laboratory notebook."[7] That is, politicians routinely ignore the truth of $2 + 2 = 4$ in favor of a more politically acceptable perception, such as $2 + 2 = 5$, and so the policy response is doomed to failure.

Another version is that many alleged "experts" are in fact charlatans in the Dr. Noitall mold: "Too often, the 'experts' who now come forward are pseudoscientists. . . . They play to the six-second soundbite, and they do it well: Their brand of 'yellow science' is remarkably effective in energizing people and motivating politicians."[8] A similar interpretation blames "seemingly qualified and objective medical and scientific professionals prostituting their credentials in . . . attempts to serve their real political agendas."[9] Overall, politicians and the voting public are alleged to be too scientifically illiterate, or too intellectually corrupt, to distinguish between the real experts and the fakes.

A more benign view—and perhaps the one most commonly held by scientists—is that the problem is merely one of public education: "[We] will not have effective political decisions unless our leaders and the voting public can better cope with scientific and technological concepts. . . . A scientifically educated citizenry and a concerned scientific community cannot remain just a desirable goal for our country. Increasingly, as we face technological accidents of global scope, the hole in the ozone layer, the terrifying global warming trend and so many other issues, it is becoming the price of our collective survival."[10] According to this view, a scientifically literate public will elect equally literate public officials, and the consequence will be effective collaboration among policy makers, scientists, and other experts to ensure the creation of wise policies based on authoritative scientific information.

As intuitively appealing as these explanations might seem, they arise from a misdiagnosis of the problem. Although politicians often display a shocking degree of scientific illiteracy, it is rare indeed that any major policy decision hinges on the type of unmitigated nonsense symbolized by Dr. Noitall's "2 + 2 = 5." This is not to say that politicians never make nonsensical decisions, but such nonsense rarely derives from a failure to accept the advice of scientists. The assertion that controversial policy problems will be wisely resolved as soon as politicians accept the value of scientific "truth" over public "perception" can be extrapolated neither from the operations of the political arena nor of the scientific laboratory. In fact, scientific information and expertise are often intrinsically unsuitable for arbitrating or resolving political controversies in a democratic society.[11]

In order to understand why this must be so, it is necessary to consider briefly the nature of political controversy* and the ways in which science can respond to such controversy. One useful description suggests that politics "is the process by which the irrational bases of society are brought out into the open. . . . [It] is the transition between one unchallenged consensus and the next. It begins in conflict and ends in a solution. But the solution is not the 'rationally best' solution, but the emotionally satisfactory one."[12] On the surface, this perspective might suggest that politics would always benefit from scientific input because such input would point toward a "rationally best" solution. In reality, when authoritative scientific answers to policy questions are available, then the questions are not controversial and there is no political conflict.[13]

The idealized—or mythological—value of scientific expertise in the political arena is its authoritativeness—its capacity to tell us what is "rationally best." Yet as we shall see, the existence of political conflict is a virtual presupposition of scientific controversy. And, in the absence of scientific consensus, there will always be legitimate scientific experts available to support the opposing sides in any conflict. The capacity of scientific expertise to contribute to dispute resolution is therefore negated, and claims to authoritativeness must collapse, as political adversaries call upon highly credentialed and well-respected experts to bolster conflicting political positions. It is mutually assured self-destruction.[14]

Furthermore, the expectation that political solutions can or should await the achievement of scientific consensus is usually unreasonable. The types of broad political or social problems that may appear most

*Here, as elsewhere in this chapter, political controversy is meant to include only those disputes that are underlain by, or related to, scientific or technological issues.

especially to cry out for scientific guidance are in fact the least likely to benefit from this guidance. Whereas an "emotionally satisfactory" political solution to such problems requires a working consensus—a sufficient public and political majority to stimulate action and exercise political influence over outcomes—scientific consensus requires virtual unanimity among acknowledged experts in order to avoid the self-negating process of dueling scientific experts. In the case of highly complex and comprehensive social issues, scientific consensus over policy-relevant questions is rarely achieved on a time scale even of a decade or two—as shown by the ongoing debate over the regulation of pesticides and other toxic chemicals—and indeed such consensus may never be achieved at all. Political action, in contrast, must often be taken more rapidly, both to forestall uncertain but conceivable consequences and to meet the responsibilities of representative democracy. For example, the use of toxic pesticides is regulated by the government even in the face of a widening scientific debate about how to measure the impact of pesticide use on human health.[15]

Whereas the myth of authoritative science suggests that scientific input can provide a rational basis for forging political consensus by separating "fact" from "perception" (if only politicians and voters were educated enough to tell the difference), in practice the converse is generally the case: political controversy seems uniformly to inflame and deepen scientific controversy. There are several reasons why this must be so. One is that the information needed by politicians to resolve major conflicts is precisely the type of information that scientists are least likely to deliver authoritatively: prediction of the future. As politicians debate the need for a new law or the design of a new program, they are actually debating the expected outcome of this law or program and its future impact on society: "Decision making is forward looking, formulating alternative courses of action extending into the future, and selecting among the alternatives by expectations of how things will turn out."[16] For science to make a decisive impact on politics, it must enhance the "forward looking" ability of politicians. And although the ultimate test of scientific knowledge may be its predictive capability, this capability only rarely—and perhaps never—achieves a scientific consensus in areas that are useful to policy makers in the midst of controversy. It is one thing to predict the existence of a subatomic particle or the conductive properties of a material or even the behavioral response of a human to a specific stimulus. It is quite another to predict successfully the outcome of the far more complex interactions that motivate major political controversies, such as local and regional variations in the earth's climate over the next century or the supply and demand curves for petroleum in the United States over a similar time period.[17]

A second reason for the synergy between political and scientific controversy is that political debate significantly raises the stakes on being scientifically "right." If a field of research makes the transition from an academic issue to one that is politically "hot," then scientific uncertainties suddenly take on political significance, the incentives to air scientific disagreements publicly are greatly amplified, new research may be undertaken that reveals new uncertainties or questions, and scrutiny of scientific results by both professionals and laypersons increases. Consensus is therefore much more difficult to achieve than for an issue that nobody outside of the laboratory cares much about. The vicious scientific dispute that arose when DNA "fingerprinting" was brought out of the laboratory and into the legal system, discussed below in more detail, is an example of this type of snowballing controversy.

Third, scientists themselves may have a political, intellectual, or economic stake in the outcome of a political controversy, and they may therefore interpret their scientific information in a way that favors their own predisposition. Even without such a predisposition, any scientific uncertainty is bound to be exploited by the political players on opposing sides of a controversy to help legitimate their position. This process was starkly illustrated when scientists from the Exxon Corporation and the federal government's National Oceanic and Atmospheric Administration aired the results of independent research into the environmental impact of the 1989 *Exxon Valdez* oil spill. It can come as no surprise that the scientists hired by Exxon "claimed that the massive spill from its tanker . . . has had little lasting effect on the wildlife of [Alaska's] Prince William Sound,"[18] even as the government researchers determined "that the Sound was still staggering from a major ecological blow."[19]

Scientists cannot even agree on standards for assessing the legitimacy of scientific information. The 1993 Supreme Court case *Daubert v. Merrell*

Dow Pharmaceuticals, which concerned birth defects allegedly caused by the drug Bendectin, hinged on this very issue: What criteria should courts use in determining the admissibility of scientific evidence? How rigid should such criteria be? Must this evidence be "generally accepted" by the research community? Is peer review always necessary? Is publication of results? Or is the professional competence of the researcher and conformance to accepted laboratory practice a sufficient standard? Both plaintiff and defendant were able to mobilize an impressive array of renowned members of the research community to support their own position. Twenty-two amicus briefs were filed in the case, most of them addressing philosophical questions related to scientific standards of proof.[20] If the scientific community cannot agree even upon the fundamental question of what constitutes valid data and conclusions, is it reasonable to expect that the rest of the world can do so?*

Authoritative scientific advice is least likely to be available when it is most needed. Scientific experts do not speak with a unified voice when issues are scientifically and politically controversial. Societal problems cannot be placed in storage while scientists grapple for a better understanding of underlying technical issues. Scientific insight into the origins of a problem does not automatically translate into wise guidance for political or social action aimed at addressing the problem. Excessive dependence on scientific advice therefore may impede the democratic resolution of societal problems. These points can best be illustrated by taking a closer look at some specific examples.

DNA Fingerprinting: Disorder in the Court

One particularly interesting—and entertaining—controversy over the application of scientific knowledge to societal problems is the dispute over DNA "fingerprinting." This forensic technique can be used to compare

*This point in no way implies that science is indistinguishable from nonscience or pseudoscience (e.g., creationism or astrology), which are neither based on observational data nor amenable to evaluation and modification using the scientific method. The problem, instead, is whether "general acceptance" of a scientific argument by the relevant community of experts is an appropriate criterion for assessing the validity of the argument and whether arguments that do not meet this standard should be excluded as legal evidence. The court held that such a criterion is too restrictive.

the characteristics of DNA extracted from blood or semen found at the scene of a crime with samples taken directly from a criminal suspect. If the DNA samples match, then, according to proponents of the technique, they must have come from the same individual because the probabilities of a false match are vanishingly small—no greater than about one in a million and perhaps much less. The technique holds great promise as a tool for prosecutors because it offers a scientifically authoritative method for tying a suspect to a crime without the need for eyewitness accounts that are often unreliable or equivocal.

For several years before the O. J. Simpson murder trial brought DNA fingerprinting into the public consciousness, scientific controversy raged over the reliability of the technique. The nature of this controversy illuminates both the political assumptions that scientists make when they claim authoritativeness and the intrinsic elusiveness of scientific consensus when the political or social stakes are high.

The storm over DNA fingerprinting broke in a 1991 issue of *Science* magazine. This issue contained a scholarly, peer-reviewed article by two respected population geneticists, Richard C. Lewontin and Daniel L. Hartl.[21] Their paper raised the possibility of "serious errors" in DNA-matching statistics. These errors could arise, the authors suggested, from the assumption—used in DNA fingerprinting—that large ethnic groups such as Caucasians, blacks, and Hispanics are genetically "homogeneous" populations that "undergo random mating" among themselves. Based in part on this assumption, the odds of a false match between any two randomly chosen individuals from the same ethnic group are extremely low. Lewontin and Hartl suggested, however, that large ethnic groups must in fact be viewed as composites of smaller, relatively discrete "subgroups." (Caucasians, for example, can be divided into such subgroups as Italian, Irish, Slavic, and Swedish.) The genetic characteristics of individual subgroups may vary significantly from the averaged composition of the large group; characteristics that are rare in the large group could be common in the suspect's subgroup or vice versa. As a consequence, the authors argue, the technique—which ignores the internal substructure of the large ethnic groups—could be tens or hundreds of times less reliable than commonly averred by scientists testifying in court.

Even before the article was published, it ignited controversy. An Assistant United States Attorney telephoned Hartl and asked him to withdraw the article prior to its publication, claiming that it would have an adverse impact on the government's ability to prosecute cases. Respected researchers who were also scientific advocates of DNA fingerprinting lobbied the editors of *Science* not to publish the original Lewontin and Hartl article without also publishing a rebuttal. The editor-in-chief acceded to this demand, even though the original article had undergone the same peer review process as all *Science* articles and in spite of the fact that *Science* had never before published a technical article and rebuttal in the same issue.[22] *New Scientist* magazine suggested that *Science* had yielded to FBI pressure in agreeing to publish the rebuttal at the same time as the original article, and *Science* in turn threatened to sue *New Scientist* for libel, a move that one observer said was "virtually unheard of" within the publication industry.[23]

Subsequent debate was equally vigorous. Many but not all of the exchanges were highly technical in nature. The nontechnical arguments illustrated the level of emotion involved in the scientific dispute. One researcher wrote: "At the end of their article, Lewontin and Hartl make some recommendations, under the heading, 'What is to be done?' (This, you will remember, is the title of Lenin's famous 1902 pamphlet, in which he made recommendations about the future of Russia. And we all know what came of *that*.)"[24] Although one presumes that this comment was facetious, its intent was to score points in a debate that was allegedly about science.

Another scientist sought blame for the controversy outside the research system. "Ultimately at fault for the furor is our legal system, which is innocent of quantitative thinking, as well as logic, in its treatment of evidence."[25] This is a curious statement because the "furor" was sparked and perpetuated by scientific disagreement between researchers of high caliber and standing. The assertion is consistent, however, with a tendency, documented throughout this book, for researchers to view science as completely isolated from, and unrelated to, the failings of society as a whole. The suggestion that the legal system is "innocent of" logic is also somewhat baffling since legal reasoning and rules of evidence, in their essence, are systems of logic. However, they are not systems of scientific logic, and the DNA controversy illustrates why this must be so: judicial decisions must be made in the face of scientific controversy. While the jury is still out on the statistical validity of DNA fingerprinting, real juries must—and do—deliver verdicts every day. Thus, some courts have found DNA evidence to be admissible, others have ruled against its admission; some juries have found it compelling, others have not.[26]

In the struggle for scientific and political legitimacy, each side of the DNA dispute claimed to represent the consensus position. More than a year after the appearance of Lewontin and Hartl's paper, an article in *Science* stated that "there is indeed a consensus supporting the reliability of estimates of genotype probability."[27]

Lewontin and Hartl responded: "Devlin *et al.* [the *Science* article authors] assert a 'consensus' favoring the multiplication rule for estimating genotype probability, but provide no supporting evidence or documentation. As it happens, an informal telephone survey by the population geneticist Charles Taylor of 33 population geneticists . . . found only 11 (33%) supportive of the method . . . 19 (58%) critical of the method, and the remaining three (9%) uncommitted. It seems that on the basis of this survey there is, indeed, a consensus in the sense of Devlin *et al.*, but not in the direction they say."[28]

Devlin *et al.* responded in turn and not without some small attempt at humor: "We stand by our claim of consensus within the relevant scientific community: the Taylor-made 'informal telephone survey,' mentioned by Hartl and Lewontin, being neither random nor unbiased, would not be taken seriously by statisticians and has had no effect on admissibility decisions [by the courts]."[29]

These conflicting claims of consensus are especially interesting because, as scientists never tire of reminding the laity, scientific truth is not established by popular vote. At its height, the dispute over DNA fingerprinting admitted of no basis for concord on either scientific or philosophical grounds. One researcher suggested: "The vehemence and lack of scientific objectivity that appear to surround this issue indicate that there may be important concerns other than scientific ones."[30] Yet each side of the debate clearly believed that it was supported by "scientific objectivity." Furthermore, as exposure to any scientific journal or professional meeting will confirm, "vehemence" in scientific debate is no uncommon thing, even if the debate is utterly devoid of societal implications. The allegation that objectivity in the DNA controversy was sacrificed to "important concerns other than scientific ones" implies that participants in the debate failed to adhere to the objectivity of the laboratory. A more useful interpretation might suggest that different scientific viewpoints represent an inextricable mix of objective science and subjective values filling the gaps where data are equivocal. After all, both sides of the DNA debate had access to the same theories and data. But the frames of reference within which they viewed those data reflected such human attributes as experience, political and philosophical leanings, ego, pride, and concerns as parochial as institutional affiliation and economic self-interest. These different reference frames are not easily or quickly reconciled by more data or more argument. Thus, one observer of the controversy suggested that "the debate is not about [scientific] right and wrong but about different standards of proof, with the purists on one side demanding scientific accuracy and the technologists on the other saying approximations are good enough."[31] Or, as a geneticist explained: "This is a religious argument. . . . We are talking about matters of faith that are not likely to be settled by reason, which is why they are at each other's throats."[32]

How, then, can society—which has a not insignificant stake in the outcome of the dispute—take action in the face of this scientific controversy? The debate is highly technical, and the experts have fundamental disagreements that they cannot settle. Both sides continue to declare victory and to attack each other with undiminished fervor.[33] Meanwhile, tighter laboratory standards and more conservative statistical assumptions have increased the reliability of DNA fingerprinting by sidestepping rather than resolving the scientific debate. Yet, because this debate rages on, legal ambiguity and controversy cannot be entirely allayed, and scientists will keep battling it out not only on the pages of technical journals but as expert witnesses in the courtroom as well.

Global Climate Change: An Atmosphere of Uncertainty

The dispute over DNA fingerprinting is focused on a particular scientific problem with particular societal implications. The policy question is clear, and it coincides with the scientific question: How reliable is the technique as currently used? In contrast, global climate change presents a comprehensive array of scientific questions and policy challenges with broad implications for society. Global change is also frequently mentioned as a prime example of a policy problem that will only be solved with the benefit of authoritative scientific input.[34]

For many years, scientific research has suggested that the accumulation of carbon dioxide and other "greenhouse gases" in the earth's atmosphere could lead to rising average global temperatures.[35] Although many of these gases occur naturally, they are also being produced and released into the air in ever-increasing quantities by human activities, such as the combustion of hydrocarbons in automobiles, factories, and power plants. Greenhouse gases cause the earth's atmosphere to trap radiant heat that would otherwise radiate outward into space. Some trapping of this radiation is natural and necessary to maintain temperatures that allow life to flourish, but significant increases in the amount of greenhouse gases could cause increased trapping and lead to major changes in the earth's climate. In addition to rising temperatures, possible consequences of an increased greenhouse effect include bigger and more frequent storms, droughts, and other extreme meteorological events; greater variation in weather conditions between and within regions; and melting of polar ice caps, leading to a rising sea level. The cultural implications of these changes include dislocation of populations and destruction of property due to more violent and frequent natural disasters, disruption of global agricultural production, and flooding of coastal zones that include many of the world's great cities.

Public concern in the United States over global climate change was galvanized in the late 1980s by a series of record-hot summers and high-profile congressional hearings where well-respected scientists predicted rising global average temperatures over the next several decades.[36] In response to this concern, the federal government initiated the United States Global Change Research Program (USGCRP) and assigned it a variety of responsibilities: to investigate global warming and other possible consequences of climate change; reduce scientific uncertainties regarding future climatic conditions; increase scientific understanding of atmospheric, oceanic, and earth processes; and provide information that policy makers could use in responding to the threat of climate change.

The USGCRP is designed to address a staggering range of research questions and problems. It will call upon scientists in a variety of disciplines using state-of-the-art technologies ranging from supercomputers to an array of climate-monitoring satellites. One government summary of the program lists 153 projects in thirty-five major areas of research under seven areas of "science priorities." The ultimate goal of this effort is: "An improved predictive understanding of the integrated Earth system, including human interactions, [that can] provide direct benefits by anticipating and planning for impacts on commerce, agriculture, energy, resource utilization and human safety."[37]

Global climate change is a politically controversial issue and will remain so. The human contributions to climate change are rooted in the very essence of industrial and technological society: generating power, driving cars, clearing land, operating factories, heating and cooling homes and buildings. Actions—even relatively modest ones, such as increased energy taxes—aimed at reducing the emission of greenhouse gases invariably encounter fierce political and economic opposition. As one member of Congress warned, efforts to reduce carbon dioxide emissions through tax measures would lead to "a potentially radical change in our way of life, complete with a crushing blow to our economy and a tax burden we pass on to the next generation."[38] Although the social and economic consequences of significant changes in global climatic patterns could be catastrophic on every scale from local to global, the lack of certainty about when these changes will take place, where and how rapidly they will occur, how severe they will be, and who will suffer has so far been sufficient to prevent any concerted political action that might perturb, even to a small degree, existing social and economic institutions.

So far, the government's principal response to the possibility of global climate change has been the support of scientific research. The policy rationale for this research is rooted firmly in the myth of authoritativeness. The essence of the USGCRP is fundamental research into the physical, chemical, and biological aspects of the "Earth system." The scientific goal of this research is "the achievement of a predictive understanding of Earth-system behavior on time scales of primary human interest."[39] This predictive understanding will "support national and international policy formulation."[40] In short, the program aims, through the accumulation of authoritative scientific knowledge and facts, to provide a foundation for the political and social response to global change, if indeed such change is occurring.

This simple formula is flawed at every level. First, the natural science research that forms the core of the program is largely divorced from the concerns and needs of policy makers. The ability of scientists to achieve consensus on aspects of global change will be directly proportional to the political irrelevance of the aspects that they are studying. The type of knowledge that the program will most effectively produce—fundamental scientific knowledge about earth-system processes—will be neither a prerequisite for nor a guide to wise social and political action. Just as the driver of a car, stopping to ask directions, would not be much helped by a sidewalk lecture on the thermodynamics of the internal combustion engine, neither will politicians find that an improved understanding of the intricacies of atmospheric processes can much help them to evaluate policy options for responding to global change.

Second, increased efforts to predict global climate change will lead to increased scientific controversy as the research community plunges more deeply into the details of the problem and politicians seek answers that will give them the freedom to satisfy their various constituencies—environmentalists who may expect immediate action, for example, or business leaders who may want no action. The hope for a predictive capability is almost certainly a vain one. Not only does modern science have no experience of success with prediction of highly complex, evolving, interactive systems but new scientific theories of complexity suggest that such systems may be inherently unpredictable in detail.[41]

Third, even if a predictive ability were achieved—even if scientists were able to agree, say, that global average temperature will rise four degrees Centigrade by 2050, rainfall in the midwestern United States will decrease by 25 percent, and sea level will rise five feet—this knowledge would merely sound the alarm with greater precision; in no way would it guide politicians and governments in their efforts to respond to the predictions. Furthermore, time frames for achieving a marked improvement in predictive capability are generally stated to be on the order of ten to thirty years.[42] One may choose to accept such estimates, but if governments refrain from taking action during the interim, they could find themselves in the year 2020 facing the certainty (rather than the mere probability) of climate change, with the time horizon for response several decades shorter than it was when the alarm was first sounded in the 1980s and with the options for incremental policy response foreclosed by the onset of crisis. Thus, the USGCRP, as currently structured, best promises to exert a formative influence on policy decisions only if global climate change turns out to be a trivial problem calling for no concerted, major governmental action.

Finally, implicit in the rationale for the USGCRP is the assumption that society can and should take no major action until scientific consensus on the extent and characteristics of global climate change has been achieved. "The scope, if not the existence, of the problem remains, after all, a matter of serious scientific inquiry," writes one critic. "Global warming may be one of those issues that does now justify another study commission, but perhaps no further intervention in the market."[43] *The Economist* seconds the opinion: "Forcing up [energy] efficiency in order to force down emissions is right only if governments are sure global warming is a grave hazard."[44] Such certainty is only likely to be achieved after the fact.

Despite these inherent, comprehensive flaws, neither politicians nor mainstream scientists are likely to blow the whistle on the current arrangement because both derive substantial benefit from it—scientists, who receive generous federal support for research, and politicians, who can claim that they are taking responsible action without having to take any political risks.

But what type of information would policy makers need to justify and motivate "intervention in the market" or other action? How will a more comprehensive scientific understanding of ocean circulation, cloud behavior, biogeochemistry, paleoclimatology, and other related areas of research enable a politician or a voter (or even a scientist) to evaluate appropriate responses and choose a course of action? Even assuming that a predictive understanding of climate change can be achieved, what policy response is implied in an authoritative prediction of a four degree Centigrade average temperature increase over the next half-century?

Another way to consider the problem is from the perspective of the policy maker. Whereas the scientist wants to understand climatic phenomena, policy makers will have to make decisions that serve their constituencies. If scientists say that the average global temperature will rise by four degrees Centigrade by the year 2050, members of Congress will want to know whether the temperature in their districts will also rise by that amount, whether local living conditions will change for better or for worse, and what actions will have to be taken in response to adverse changes. If scientists say that a 20 percent reduction of global carbon dioxide emissions will prevent significant melting of polar ice caps, the Secretary of State will want to know how to forge an international agreement to reduce emissions without destroying global trade or regional economic stability. The scientific questions focus on how, why, and when the climate will change. The policy questions are entirely different; they focus on understanding and finding ways to mitigate or respond to the negative impact of climate change on society. Yet policy action has focused almost exclusively on funding research aimed at scientific questions while largely ignoring much more difficult questions of policy. This course of action reflects the desire of politicians to postpone difficult decisions, the belief that a better scientific understanding of natural processes will make those decisions more self-evident and less painful, and the hope that the global climate problem will, if given enough time, just go away.

Given these fundamental problems, how, then, should society respond to the threat of global climate change under ongoing conditions of scientific uncertainty? Although the USGCRP treats climate change as a grand, comprehensive research problem that will someday yield an authoritative, comprehensible explanation, in fact global change implies a mélange of hundreds of smaller problems, such as: the response of existing agricultural systems to different types of climatic conditions and different rates of climate change; the potential for reducing carbon dioxide emission through different types of tax and technology policies; the regulation of development near shorelines; the migration of populations displaced by drought, flooding, desertification, and other disasters; the phasing out of greenhouse gases such as chlorofluorocarbons; the development of new and better sources of renewable energy; the control of large-scale deforestation. When reduced to such components, global change looks rather familiar, perhaps painfully so. But the possibility of response is less daunting, the mechanisms for response are conceivable, the excuses for avoiding response are less compelling, and the consequences of error are less serious and more easily correctable. In unifying such problems under the umbrella of global change research, however, the promise of achieving a comprehensive, predictive understanding of the "Earth system" in the future may have the paradoxical effect of undermining the incentives for political action, even at a modest level, in the present.

Meanwhile, the global change research agenda concentrates almost entirely on large-scale atmospheric, oceanic, and earth processes. A trivial 1.5 percent of the research budget is devoted to understanding the societal impacts of and human response to climate change. Even less has been allocated to exploring mechanisms for bridging the gap between the information needs of policy makers and the research agenda of scientists. As a consequence, policy makers will find themselves increasingly frustrated by the inability of scientists to provide definitive guidance on global change policy, and scientists will find themselves increasingly exasperated by the scientific illiteracy of politicians who want authoritative answers to seemingly simple questions. Meaningful political and social action will continue to be held hostage to these proceedings.[45]

Stratospheric Ozone Depletion: Getting It Right (by Accident)

The case of stratospheric ozone depletion illustrates a science-policy success, in that a convergence of scientific and political consensus seems to have led to decisive and timely political action. The path to consensus, however, was indirect and somewhat fortuitous; the contribution of public opinion to a satisfactory policy solution was as important as that of scientific expertise. Early warnings about depletion of the earth's ozone layer, which helps to filter out the sun's ultraviolet radiation, came in the late 1960s as the United States was planning to build its first supersonic transport (SST). Preliminary research suggested that exhaust from the SST could contribute to ozone destruction, and members of the President's Science Advisory Council (PSAC) advised President Nixon against construction of the aircraft. Nixon, disenchanted with this advice, as well as advice and opposition he had been receiving on other issues such as the antiballistic missile system, disbanded PSAC in 1972 and eliminated the position of presidential science advisor.[46]

In response to public concern over the SST controversy and air pollution in general, Congress provided new funding for research in the atmospheric sciences. In 1974, chlorofluorocarbons (CFCs)—a class of chemicals used as coolants in refrigerators and air conditioners, propellants in aerosol sprays, and solvents in the electronics industry—were identified by researchers as a potential cause of stratospheric ozone depletion. Energetic scientific controversy ensued. The economic stakes were high because of the importance of CFCs to the chemical industry, which marshaled its considerable scientific expertise to argue against government restriction of CFC use.[47]

By the late 1970s, American consumers, concerned about the environmental implications of CFCs, voluntarily cut their purchases of aerosol sprays by two-thirds. This collective and relatively spontaneous action was taken amidst ongoing scientific controversy and a notable absence of any authoritative voice about the environmental effects of CFC use. In the absence of concerted international action, consumer activism was also largely a symbolic gesture. But politicians understand the importance of symbolism (sometimes preferring it to substance), and in 1977 the U.S. Congress—always willing to make a virtue out of reality—authorized the government regulation of CFC consumption that led to the 1978 ban on "nonessential" CFC uses, such as in aerosol sprays.[48]

In the early 1980s, urgency over the ozone problem temporarily diminished. New research suggested that the threat of ozone depletion was probably not serious. International negotiations over control of CFC emissions bogged down. The United States continued to push for an international ban of CFCs in aerosols, in part because it would offer American manufacturers a competitive advantage over foreign companies that had not yet phased out CFCs. European nations, most of which still had a significant economic investment in CFC-propelled sprays, continued to argue that CFCs were not harmful to the ozone layer.[49]

In the mid-1980s the scientific momentum shifted again as additional research suddenly began to point once more toward the potential for serious CFC-induced ozone depletion. Predictive models showed that significant, measurable depletion would occur by the middle of the twenty-first century. International negotiators, responding to this growing concern in the scientific community, crafted the Montreal Protocol, an agreement that called for a 50 percent phase-out of CFC consumption by the industrialized nations before the year 2000. Twenty-four countries signed the agreement in 1987. The Protocol was viewed "not [as] a response to harmful developments or events, but rather [as] *preventative* action on a global scale."[50]

While the Montreal Protocol was being negotiated, scientists first discovered the existence of a "hole" in the ozone layer above the Antarctic. The chief U.S. negotiator for the Protocol has emphasized that this discovery had little effect on the international negotiations because the research was preliminary and because the ozone hole could not, at first, be connected to CFC use. Additional study, however, confirmed the connection and suggested that serious, global-scale ozone depletion had already occurred, half a century earlier than scientists had anticipated: "The [scientific] models on which the Montreal Protocol was based had proved incapable of predicting either the chlorine-induced Antarctic phenomenon or the extent of ozone depletion elsewhere."[51] The Protocol turned out to be a "response to harmful developments" after all. In 1990, the protocol was strengthened with the aim of achieving a 100 percent CFC phase-out by the end of the century. By 1992, 68 nations had signed on.[52]

In 1993, the process that began twenty years earlier when President Nixon fired his science advisors came full circle: President Clinton fired the director of energy research at the Department of Energy for publicly maintaining that the effects of ozone depletion were being overemphasized by the government and that more research was required before additional federal policies could be developed. The locus of scientific dispute had shifted from establishing the fact of ozone depletion to determining the implications of that fact for society.[53]

The case of ozone and CFCs illustrates one way in which science advice may effectively serve the political process, as growth of scientific consensus about ozone depletion was accompanied by increasingly stringent controls on CFC consumption. Yet there are many troubling aspects of this story. Although President Nixon's disbanding of PSAC in 1972 represents the type of political disregard for scientific advice that infuriates both scientists and a concerned public, the scientific basis for PSAC's original warning about destruction of the ozone layer by the SST was probably wrong. Although Congress banned CFCs in aerosols in the late 1970s, this ban was politically and economically feasible because the public, which had already cut down on its consumption of aerosols, supported such action even in the face of considerable scientific uncertainty. Although the United States pushed strongly for an international ban on CFCs in the early 1980s, they did so even as researchers were minimizing the risk of ozone depletion. Although the Montreal Protocol represented a responsible international reaction to a growing scientific consensus, the consensus was based on an inadequate understanding of ozone depletion, including a failed predictive capability. Although the discovery of the Antarctic ozone hole and subsequent research demonstrated the connection between CFC use and ozone depletion, the scientific debate then turned to the effects of this depletion and to another presidential firing.

If humanity has successfully addressed the problem of stratospheric ozone depletion, at least for the time being, then this outcome grew out of a fortuitous confluence of scientific controversy, politics, economics, and international diplomacy. The wisdom of the CFC ban is now greatly reinforced by the discovery of the ozone hole, a type of observational smoking gun that very rarely saves the day during political disputes. Yet the ozone hole was discovered during a time of growing policy consensus. Perhaps the scientific consensus that emerged after this discovery would have been far more fragile if the policy consensus had not already existed. Indeed, a minor political backlash against the CFC ban has already occurred.[54] Should policy makers choose to revisit the issue, the remaining scientific uncertainties that surround the ozone depletion problem will be quite sufficient to stimulate continued political debate—and continued research—for many years.

Authoritative Politics

Science, and science advice, more often lead to political controversy than to political accord, at least in the short term. This does not demean the potential contribution of scientific expertise to the governing process: political controversy, after all, is the lifeblood of democracy. If, however, our political institutions were consistently to defer action in the expectation that authoritative scientific guidance would resolve political dispute, then action might never be taken; scientific research would become a political alternative to genuine social reform. To some extent, this alternative has been exercised in the case of global climate change.

A famous political axiom suggests that "every nation has the government it deserves." So, perhaps, does every government have the science it deserves. As unsatisfactory is it might seem, the science that is chosen to inform policy decisions will usually be a reflection of society's prevailing cultural values and political milieu. Thus President Bush and President Clinton both claimed to be advocates of wetland preservation, but they used different "scientific" definitions of wetlands to support different policies for preservation. Any policy maker can choose from a generous menu of options in seeking scientific support for political positions. Some biologists believe that global biodiversity is threatened by environmental degradation; others claim that the problem is exaggerated. Some energy experts view renewable energy technologies such as photovoltaic cells and wind-driven turbines as potentially significant sources of power in the near future; others argue that they will be trivial. Some toxicologists feel that dioxin and environmental estrogens are major threats to the health of humans and animals; others are convinced that the threats are not serious. Some hydrologists believe that groundwater pollution represents a major potential threat to human health; others feel that the costs of cleanup far outweigh the benefits. Some epidemiologists are greatly concerned about the human health effects of electromagnetic fields; others believe the problem is overblown and perhaps even nonexistent.[55]

More scientific advice, and more data delivered to the government, will not soon resolve most of these disputes. In fact, closer links between researchers and policy makers could exacerbate some problems by adding new layers of controversy and complexity to public debate, obfuscating the political or cultural roots of social problems, and alienating nonexpert participants in the political process. Nor will a more scientifically literate populace necessarily contribute to faster or more satisfactory dispute resolution—more literate members of the public will presumably give more credence to scientific information that supports their preexisting political inclinations. Overall, a dependence on the ability of authoritative science to rationally adjudicate major political conflict probably heightens the level of both political and scientific controversy. Politics, after all, is mostly about resolving disputes. Science, on the other hand, is a process of inquiry, of asking questions. When a field of inquiry stops yielding new questions, it is no longer science. Definitive answers are uncommon indeed, and elegant predictive descriptions of entire systems are the rare stuff of scientific revolutions. Research in the natural sciences is therefore an effective tool for alerting society to potential problems, but it is intrinsically ill suited for prescribing solutions to those problems.

These arguments do not mean that scientific research related to politically controversial subjects should not be undertaken or vigorously pursued or that such research cannot have a useful role in government. They do suggest that efforts to translate scientific information directly into wise or sensible or "correct" policy decisions will likely be doomed to failure. Research in the natural sciences related to such major problems as global change should be viewed not as a prerequisite for action but as a process that can help constrain the terms of political debate and delineate boundaries within which policy decisions may make sense. For example, a policy response to global climate change that assumes a ten degree Centigrade average temperature increase, or any significant decrease, over the next several decades lies outside the boundaries of the scientific dispute. But within these boundaries, there are an infinite number of policy options, and choosing between such options will inevitably be a political process, a reflection of the character of the human culture, rather than the prescriptions of authoritative science.

Science can contribute basic facts that are relatively noncontroversial components of broader issues. Thus, there is little disagreement within the scientific community about the theoretical basis for the greenhouse effect, or about the contribution of carbon dioxide to that effect, or about the increasing levels of carbon dioxide in the atmosphere. There is even a growing consensus that a modest global-scale warming trend documented by climate researchers for the twentieth century does indeed result from anthropogenic emission of greenhouse gases, rather than natural climatic variations.[56] These facts justify the consideration of global climate change from a political perspective, but they do not amount to a useful predictive capability, nor can they be translated into a particular course of wise political action and societal response.

Science may also define or illuminate new realms of ethical debate by enlarging society's worldview. The concepts of sustainability and inter-generational equity flow in part from scientific inquiry into global climate change and other long-term environmental impacts of industrial society. The possibility that economic development and resource utilization decisions today may influence the capacity of the planet to support human society in the future introduces a new moral dimension into policy dialogue. This dimension may even feed back into the research system and stimulate new areas of investigation. The sustainability concept, for example, has had a considerable impact on the research agenda of such diverse fields as agriculture, energy supply, economics, and biology.

In the end, the most authoritative component of political debate is not science but the matrix of cultural values that guides society in its struggle to advance. The scientific information that politicians use to aid their decision making is explicitly chosen and interpreted—with the help of scientific experts—to support action that is consistent with those values. In this way, technical data can become a surrogate for values, and the real terms of debate may be concealed or confused. Scientists express anguish and contempt as they watch politicians ignore "the teachings of science" in favor of "the preaching of moral values."[57] Yet one might reasonably infer that the true source of this anguish does, in fact, lie in the moral realm. One would not expect scientific experts to support policy options that are scientifically rational but morally or politically repugnant to them as human beings. But neither does one expect to hear scientists say, "this data or theory is particularly compelling to me because it implies a course of political action that I find ethically or politically acceptable." If they did, the special legitimacy that comes with scientific expertise would be sacrificed; authoritativeness would vanish. Nothing would remain but a knowledgeable human being with an informed opinion.

The Dr. Noitall quotation at the beginning of this chapter fails to characterize the impact of science on policy making because it asserts that most politicians and voters are too ignorant or venal to care about the difference between "truth" and "perception." The great danger of this perspective is that it would have us replace the collective conscience of a nation with the equally subjective conscience of the scientist. In all likelihood, most people would be absolutely delighted to know the "2 + 2 = 4" of climate change, ozone, DNA fingerprinting, and toxic chemical use, if such certainties existed. But they usually don't. In the absence of authoritative science, we are left only with our political, imperfect selves upon which to depend.

The Myth of the Endless Frontier

Where man is not, nature is barren.—William Blake, *The Marriage of Heaven and Hell*

ALTHOUGH THE GOVERNMENT does support research in the expectation that new knowledge can guide policy making, and thus help indirectly to address a range of societal concerns, a much more significant motivation for federal sponsorship of R&D is the belief that many problems facing humanity can be directly confronted with concrete products and processes created in the laboratory. The effort to transform scientific ideas into tools for the direct resolution of societal problems is given shape through a view of technological innovation called the "linear model." According to this model, the path from fundamental scientific research to useful products is an orderly progression, starting with the creation of new knowledge in the basic research laboratory and moving sequentially through the search for applications, the development of specific products, and the introduction of these products into society through standard commercial channels or through government programs such as national defense. Like the god of deism, the federal government has merely to provide adequate support for basic research in order to set the whole sequence in motion.

As an intellectual construct, the linear model has been repudiated by scholars of innovation.[1] Not only are such categories as "basic" and "applied" research often artificial or arbitrary, but science and technology in the modern world are entirely symbiotic—with each other, of course, and also with economics, politics, and culture. As a practical matter, however, the linear model forms the organizational basis of the post–World War II federal R&D system—institutionalized in agencies such as the National Science Foundation and National Institutes of Health—and its metaphorical power still influences the thinking and the rhetoric of both policy makers and natural scientists: "Though times be hard, if we proceed, as our predecessors did, to sow the seed corn of research with no goal except the understanding of nature, we will ensure the harvest that will sustain our grandchildren."[2]

The continued influence of the linear model may partly reflect a hierarchical perspective in the science community that ascribes the greatest intellectual and social prestige to basic or "pure" research—the source of new knowledge—while viewing the role of applied research and technology development as more concrete, less difficult, and therefore less intrinsically worthy.[3] A crucial aspect of this line of thinking is the idea that fundamental scientific knowledge is a thing apart, accumulating as if in a reservoir, from which it can later be drawn by applied scientists and engineers who are interested not in the laws of nature but in the creation of products and processes.

In other words, in the world described by the linear model of innovation, the process of knowledge creation is intrinsically unrelated to the ways in which society may choose to make use of that knowledge: "Science enables Technology, which opens up enormous and rapidly growing ranges of human opportunities. . . . We all agree that Technology is influenced by political, societal, and commercial forces, but don't blame the drivel of American television programming on us scientists, nor our firm but flavorless tomatoes, nor our preference for Toyotas to Fords, nor the AIDS epidemic, nor the ozone hole, nor the threat of nuclear winter, nor the lack of female mathematicians and of black astrophysicists. These are real problems, but they are more societal than scientific."[4] From this perspective, society endows science with value—economic, social, cultural, and moral—by the choices it makes in assimilating and applying the new knowledge that it acquires: "The failures of our society that lead to the disasters and potential disasters facing our civilization are not the consequence of achieving the goals of science. They are the results of the misapplication of science. The goals of science are clear: in broad terms they are the basic understanding of nature and its laws."[5]

The endless frontier is the mythical territory where nature and its laws await their gradual and progressive discovery. Scientists are the explorers venturing into this unknown terrain, and scientific theories and data are the products of their exploration. The myth of the endless frontier asserts that new scientific knowledge is therefore nothing more than the passive and disembodied expression of humanity's desire to better understand nature. At the time of its discovery, scientific knowledge is thus an amoral entity—indeed, "premoral" might better describe its relation to human values because new knowledge reflects nothing but the structure of nature even though it may later be used by humans to pursue the subjective goals of society. The linear model, which describes this later application of scientific knowledge to human problems, can therefore be seen not simply as an explanation of how scientific research leads to technological innovation but also as an implicit description of the process by which scientific knowledge begins to accrue social meaning.

Yet a historical perspective on the rise of modern scientific thinking suggests that the moral separation of scientific knowledge from its uses in society is less complete than the myth of the endless frontier and the linear model of innovation might lead us to believe. The historian Lynn White has argued that the acceleration of both scientific and technological progress in Europe during the late Middle Ages has its roots in Christian morality.[6] Other cultures had produced marvelous science and marvelous technologies, but never had they done so simultaneously with the ardor and genius shown by Renaissance Europe. The European revolution in science and technology flowered in a world where one's work and one's life were inseparable from Christian morality. The great European scientists, among them Newton, Kepler, and Galileo, were inspired in their quests to better comprehend nature by a desire to know and comprehend God and God's works: "From the thirteenth century onward into the eighteenth, every major scientist, in effect, explained his motivations in religious terms. Indeed, if Galileo had not been so expert an amateur theologian he would have got into far less trouble [with the Church]; the professionals resented his intrusion. It was not until the late eighteenth century that the hypothesis of God became unnecessary to many scientists."[7]

Technological innovation, meanwhile, became a spectacularly potent tool for carrying out the Christian God's charge that humanity exercise dominion over nature. It was this latter attribute—a spiritual mandate for technology development, in essence—that may have been uniquely Christian.[8] After all, the industrial revolution was born in Europe and not, for example, in China, even though the Chinese had a two-millennium advantage in iron-working technologies and had invented the printing press and moveable type several centuries before Gutenberg.[9] Innovation and faith were further joined as the technological supremacy of Christian Europe enabled the progressive conquest not merely of nature but of most of the rest of the pagan human world. Wealth, resources, and knowledge acquired in conquered lands motivated and accelerated further technological progress. Thus impelled by a Christian vision of both knowledge (for the glorification and comprehension of God) and progress (for achieving mastery over nature and for spreading the faith), scientific and technological advance continued to accelerate, ultimately reaching the stage where progress in one became inextricably dependent on progress in the other.

Even the philosopher Francis Bacon (1561–1626), now considered to be the father of modern secular science, accepted and worked within the Christian moral tradition. Bacon's famous dictum "Nature to be commanded must be obeyed" grew from the Christian desire both to understand God's laws in nature and to exercise dominion over nature. Bacon postulated an explicit linkage between understanding nature through scientific inquiry and making use of that understanding to subdue nature for the benefit of humanity. But his efforts to separate scientific from moral inquiry were not intended to remove science from the moral or theological realm; rather, he hoped to find a path—scientific inquiry— that would better allow humanity to serve the goals of that realm: "Only let mankind regain their rights over nature, assigned to them by the gift of God, and obtain that power, whose exercise will be governed by right reason and true religion."[10]

Bacon's influence as an early architect of the rationale and methodology of modern science was matched only by the mathematician and philosopher René Descartes (1596–1650). While each viewed science as a process for the comprehension and conquest of nature, Descartes, in postulating the duality of mind and nature, further portrayed the natural world as explicitly mechanical—nothing more than a machine, a clockwork, to be comprehended (and thereby dominated) through the study of its component parts.[11] Through this new comprehension, Descartes foresaw worldly gain in human welfare, but the ultimate goal of scientific knowledge and the technological domination of nature was comprehensible only in theological terms. The philosopher Max Oelschlaeger suggests: "The Baconian-Cartesian dream, thus, was that humankind might transcend the Fall, rise up from its sinful condition, and create a heaven on earth. Through science human beings might realize the covenant of the Old Testament and become the master of nature."[12]

By the time—350 years later—that Vannevar Bush outlined the justifications for government support of research and development, the theological context for the pursuit of scientific and technological progress had long since been stripped away by the intellectual tradition of the Enlightenment, the mechanistic reality of the industrial revolution, and the prosperity of the modern age. Bush invoked the cultural and political myth of the creation of America rather than the Christian call to know and serve God: "It has been basic United States Policy that Government should foster the opening of new frontiers. It opened the seas to clipper ships and furnished land for pioneers. Although these frontiers have more or less disappeared, the frontier of science remains."[13] As a metaphor, however, Bush's frontier plays a role similar to that of Bacon's God—the locus of all that is unknown beckoning to be better comprehended, promising benefits, always out of reach. The technological mastery of new frontiers is the proven mechanism for enforcing humanity's dominion over nature. Thus, in articulating a political rationale for the R&D system, Bush used a powerful metaphor to connect science and technology to their cultural context, just as Bacon had done centuries before. What has changed in the intervening years, of course, is the impact of science and technology on the structure of society, which is amplified and diversified almost beyond recognition.

The myth of the endless frontier derives from and depends upon its historical and cultural roots. That we are no longer aware of or willing to discuss them does not diminish their influence. What is clear, in fact, is that the gradual submergence of the Christian moral origins of the research enterprise has been matched by a gradual increase in the moral implications of scientific and technological progress for humanity. These moral implications derive merely from the fact that almost every element of modern existence is in some way influenced by this progress. From the destructive power of modern weapons to the quest for the elimination of global scourges like malaria and AIDS, science and technology—together—are experienced as moral forces by human beings. If scientific research can plausibly be described as a value-neutral or premoral activity, it is because the force of scientific progress pushes society in the direction that it wants to go anyway, or at least is willing to go. It is only when society begins to perceive a deviation between the path of its own welfare and the path of scientific progress that the moral implications of science begin to emerge and the claims of moral autonomy made on behalf of science begin to look contrived or self-serving.[14]

Through most of human history, science and technology were indeed carried out as separate and unrelated activities, but starting in about the mid-nineteenth century, the quest for scientific knowledge and the pursuit of technological innovation began rapidly to converge. The great symbol of this convergence—both its cause and its product—was the industrial revolution, which, in its later stages, linked science and technology together in the service of human productivity and economic expansion. Perhaps the culmination of this trend—the final and irrevocable fusion of science and technology—was the Manhattan Project, wherein scientists who had previously devoted their intellects to advancing the understanding of physical reality were called upon to use their expertise for the construction of a tangible technological product, the first atomic bomb. Oddly enough, then, the U.S. government began to create a vastly expanded system of publicly funded research and development rooted in the philosophical and practical separation of science and technology at exactly the moment when their inextricability could no longer be in doubt.

Endless Frontier, Finite Earth

When Vannevar Bush chose the "endless frontier" as his principal metaphor to encompass the goals of federal science policy, he was adopting a historical perspective that is no longer much in favor. As the twentieth century comes to an end, Bush's invocation of clipper ships and pioneers sounds almost quaint. Frontiers are messy places; the conquest of the American frontier, and the great prosperity that resulted for the colonizing cultures, was of course bought not merely with energy, ingenuity, and beads but with genocide, slavery, and wanton destruction of nature. The urge to explore and the urge to conquer are not so distinct, and each facilitates and motivates the other, as Francis Bacon understood when he claimed, with great foresight, that the understanding of nature was a necessary prerequisite to its conquest.

Scientific exploration and technological conquest of the frontier of knowledge have made possible a progressive liberation from elemental want in the industrialized world that has culminated in an average standard of living that could not have been imaginable even 150 years ago. But a parallel consequence of this material progress has been an unprecedented acceleration in the exploitation, modification, and despoliation of nature. In the context of Christian culture and the Baconian creed, this consequence is not a cost but a tangible measure of progress, the evidence of human mastery over and subjugation of nature.

The rise of environmentalism in the industrialized world marks a change in perspective, and what was once viewed as a positive indicator of the human intellectual, moral, and material progress that accompanied the advancing frontier of science is now often considered to be an unacceptable cost. Progress used to be measured by how fast we could cut down trees, fill marshes, extract minerals, burn fuel, cultivate soil, pump water, build smokestacks, pave land. Now these are commonly viewed as measures of a type of decline, and some seek to retard a momentum that we once sought to enhance. Science and technology are routinely offered as keys in the effort to effect this reversal of direction; many are confident that the problems created in part by science and technology will be solved with more science and technology. "Think about the macroengineering problem of repairing the ozone layer, should that become urgently necessary," writes Leon Lederman. "Toxic and nuclear wastes, urban air, and water pollution are socioeconomic problems that would be enormously helped by imaginative science and engineering thought."[15] The causes of these challenges—for example, such manufactured substances as chlorofluorocarbons, dioxin, and neptunium 237—are thus defined not as products of science and technology but as problems of socioeconomics. Freon doesn't destroy stratospheric ozone; people do.

From such a perspective a system of science and technology designed to create environmental problems can be converted to a system that solves them. That is, the problems themselves are treated as though they were independent of the scientific and technological progress that made them possible. This is an appealing perspective because it offers a simple way out of the environmental dilemma posed by economic growth and industrialization: more science and technology (and thus more growth and industrialization).

Bacon argued that science must proceed from the specific to the general, that the true way of "investigating and discovering truth . . . constructs its axioms from the sense and particulars, by ascending continually and gradually, till it finally arrives at the most general axioms."[16] That is, the scientific investigation of small-scale natural phenomena will create bodies of fact that can be assembled like the pieces of a puzzle to yield ever more precise and useful pictures of nature writ large. Descartes pushed the method even further in his insistence that nature was nothing more than a machine constructed of describable parts, whose operation was governed by immutable, ultimately ascertainable natural laws.

The essence of the Baconian/Cartesian method—the essence of modern science—is reductionism, the search for the component facts that can permit scientists to uncover the universal laws at the core of nature. The ability of scientists working in this theological and intellectual tradition to derive "general axioms" about nature is therefore dependent upon their ability to look further and further into the details of nature's clockwork. Thus, the Baconian/Cartesian method has the paradoxical effect of increasing, over time, the reductionist tendencies of science, as researchers look to increasingly rarefied, abstract, and small-scaled phenomena and once coherent disciplines atomize into hundreds of new subdisciplines, each striving to comprehend particular and ever more circumscribed slices of reality. The exploration of nature is thereby fractionalized, not unified, as the relations between subdisciplines—and thus between natural phenomena—become increasingly difficult to recognize and the discovery of general axioms becomes increasingly elusive.

Reductionist scientific research helped to forge the path to the postindustrial world by elucidating specific natural phenomena that could be isolated, manipulated, and applied to practical tasks. This path was incremental and unpredictable. Growth could occur in any direction. Society was a phenomenon apart from nature, although acting on nature, seeking to subdue nature. The job of science and technology was to figure out ways to overcome natural barriers to human welfare. This job was an explicit part of Christian dogma, reinforced not merely by the early philosophers of science such as Bacon and Descartes but by recent voices such as Vannevar Bush, and still recognizable in the words of today's researchers. A Harvard University physicist writes in 1993: "Let us support the truly innovative people who can teach us about the natural world and how to harness it to our benefit."[17]

The idea that humanity had a moral responsibility to act *on* nature turned out to be a highly practical approach to achieving material prosperity, and it formed the philosophical foundation upon which both science and technology could be defined and their progress measured. Had European culture not adopted this subjective worldview, it and the rest of humanity might still be mired in preindustrial civilization. Yet the fact that views about the relationship between humankind and nature underlie—albeit deeply—both our current state of scientific and technological sophistication and current levels of material well-being does not imply that such views must continue to serve humanity well for the indefinite future.

The long-term effects of growing human population combined with the expansion of industrial economies, the exploitation of natural resources, the generation of waste products, and the intentional and accidental modification of the environment are subjects of much research, analysis, and debate. These trends create the real if unproven possibility that humanity's ability to augment or even to maintain its welfare may be compromised in coming decades. If the carrying capacity of the earth—its ability to support developing civilization—is indeed finite, and if human activities threaten to exceed or reduce this capacity, then the future welfare of humanity may depend in large part on our willingness to adopt an entirely new perspective on nature.* The Christian/Baconian idea that

---

*Many scientists argue, quite correctly, that scientific research has been crucial to both the recognition and monitoring of environmental degradation. However, the fact that modern

humanity is free to act *on* nature presupposes that nature is infinitely resilient and that no amount of human activity could affect the environment in a way that could stifle desired trajectories of growth and development. But once this idea is called into question, the capacity of more science and technology to successfully confront ecologic crisis must necessarily be questioned as well.

The environment that sustains life on earth is a product of complex, ongoing, and evolving interactions between physical, chemical, and biological systems. Of these systems, the most dynamic—the most rapidly changing—is human civilization. Throughout its history, of course, humanity has engaged in the task of altering various local components of the environment to facilitate its comfort and survival—damming rivers, domesticating animals, clearing forests, plowing and sowing fields—but science and technology have incalculably accelerated this process of alteration. More significantly, they have done so by introducing into the environment what physicist and historian of science Silvan Schweber calls "genuine novelties in the universe . . . objects that never existed before"[18]—the progeny of reductionist science and technology.

Ecologic crisis is systemic; it reflects the interconnectedness of those same phenomena and components of nature that modern reductionist science seeks to isolate and manipulate. For the past century or so, science and technology have combined to introduce millions of new variables— novelties ranging from radioactive waste and plastics to antibiotics and transistors—into a system that was already inestimably complex. The new science of complexity, which is arising simultaneously from disciplines as diverse as physics, economics, artificial intelligence, and the ecological sciences, recognizes the process of "emergence," whereby apparently straightforward natural or artificial systems, governed by apparently invariant and often quite simple rules or laws—a chess board, some computer software, a weather pattern—give rise to intrinsically unpredictable and often highly surprising consequences—a brilliant new offensive ploy, a completely unexpected computer glitch, a typhoon. Complexity science

research has permitted the continued diagnosis of ecological crisis does not contain within it the implication that conventional scientific methodologies can create the appropriate intellectual and technological tools for responding to the crisis.

has profound implications for the applicability of Christian/Baconian science to ecologic crisis because the concept of emergence dictates that the specific behavior of systems cannot be predicted or comprehended through the process of determining the natural laws that govern the behavior of the individual system components. The liquidity of water cannot be predicted from the behavior of individual water molecules—liquidity "emerges" from the interaction of innumerable such molecules; life cannot be predicted from the structure of DNA; the universe as we experience it today cannot be predicted from the ephemeral phenomena and particles that existed in the first instant of the Big Bang.[19]

Because mainstream science is reductionist while the environment is an interconnected, complex system, much of modern science and technology may be intrinsically unsuited to successful confrontation of ecologic crisis. The products of science and technology—"genuine novelties in the universe"—are not isolated entities; they are variables introduced into complex natural and cultural systems. When such products are designed to address a particular, highly focused problem—to change just one of the many variables in these systems—they are also likely to stimulate surprising and unpredictable consequences beyond what they were intended to accomplish. The environmental and cultural effects of the Green Revolution are a prime example. The development of new, high-yield varieties of important food crops may have had no more complex a rationale than to boost global food production and reduce world hunger, but it has also caused severe environmental and human health impacts. These include massive and often indiscriminate production and use of pesticides and other agrochemicals, reduction of genetic diversity (and thus increased vulnerability to disease) of staple crops, and disruption of stable agrarian cultures in many parts of the developing world due to economic incentives that favor large farms over small ones. The environmental movement itself can be interpreted as an emergent consequence of the Green Revolution to the extent that environmentalism was stimulated by a growing concern about the negative ecological consequences of pesticide use, as portrayed in such early environmental works as Rachel Carson's *Silent Spring*.[20] All this from the development of a few new varieties of rice, wheat, and maize.

Such concepts are neither new nor difficult, yet they do not fit comfortably into the dominant methodologies of modern science. Most research still follows the reductionist path to understanding. But the components of nature—isolated, atomized, manipulated, individually studied and understood—do not add up to nature. In a fundamental sense, the vision of Bacon and Descartes was wrong: facts can be gathered, natural laws discovered, but these cannot necessarily be assembled to yield a true picture of nature. Thus, the momentum of much research—basic research especially—may be headed in precisely the opposite direction from that needed to understand humanity's place *in* the environment. The most prestigious, most expensive research programs continue to focus on and search for fundamental components of large, complex systems: the structure of the human genome or of the subatomic components of matter, for example. David Baltimore argues that more knowledge creates more freedom: "Freedom is the range of opportunities available to an individual—the more he has to choose from, the freer his choice. Science creates freedom by widening our range of understanding and therefore the possibilities from which we can choose."[21] But reductionist science only widens understanding by narrowing its focus; it cannot create insight into the ways that more options, more choice, more novelty, may influence the quality and character of existence at the level of an individual, a culture, an ecosystem, or a planet. To the extent that reductionist science explicitly and inevitably fails to move humanity closer to the understanding of complex systems, it may in fact have the unanticipated effect of reducing freedom by creating the false expectation that the world can be manipulated in precise and predictable ways for the benefit of humanity, based simply upon a knowledge of the minute and intricate mechanisms of nature.

Václav Havel writes: "The more thoroughly all our organs and their functions, their internal structure and the biochemical reactions that take place within them, are described, the more we seem to fail to grasp the spirit, purpose and meaning of the system that they create together and that we experience as our unique self. . . . Experts can explain anything in the objective world to us, yet we understand our own lives less and less."[22] Havel is sometimes dismissed by researchers as a fuzzy-thinking science basher,[23] but his arguments are rigorous and can be interpreted quite literally. Increasingly detailed understanding of the components of physical and biological reality does not really move us closer to an understanding of emergent phenomena such as life, consciousness, or ecosystems. However, by occupying ourselves more and more with reductionism, we necessarily move away from efforts to consider and confront the world as we actually experience it. The practical implications of this growing divide may begin to press more urgently on humanity if the ecologic crisis continues to emerge as a tangible threat to global welfare.

In a purely pragmatic vein, the Christian/Baconian tradition strongly influences the organization of the research system in ways that may undermine efforts to understand causes and potential responses to ecologic crisis. The traditional moral place of humanity outside of nature has permitted modern science largely to neglect research on human behavior in its investigations of natural phenomena. It is preposterous to imagine, however, that the environmental challenges facing humanity can possibly be addressed by increasing our knowledge of the physics, chemistry, and biology of environmental processes divorced from a commensurate advance in our understanding of humankind's interaction with and effect on those processes: "The highest priority for research is to build understanding of the processes connecting human activity and environmental change.... [An] *interdisciplinary environmental social science—a field that examines the environmental effects of the driving forces [of global change]—is not yet organized.*"[24]

But reductionist science requires a relatively rigid disciplinary organization. This organization, and the specialization of activity that it promotes, has been one of the keys to the success of modern science. Because the research system is heavily weighted toward disciplinary research—in both the natural and social sciences—intellectual and bureaucratic inertia act strongly to inhibit the growth of interdisciplinary problem solving. Interdisciplinary and multidisciplinary research is tacitly and explicitly discouraged by the highly decentralized administration of environmental research in the U.S. government and by the traditional disciplinary organization of the sciences in academia. Discipline-based research organizations will always tend to resist the creation of interdisciplinary structures that may compete with them for money, space, and talent.[25] Scientists who attempt to escape the constraints of disciplinary research often find themselves intellectually isolated, and they invariably find it difficult to obtain funding through the peer review process. (In such ways are the theological roots of modern science and technology now expressed as political, bureaucratic, and intellectual inertia.) As severe as this problem may be within the natural sciences alone, it is an even more intractable obstacle to the alliance of natural and social sciences or of the sciences with the humanities. Yet ecologic crisis, and its potential solutions, are comprehensible in terms of no particular discipline but must be viewed instead through a fuller, more synthetic understanding of the interactions of human activity with and within the physical and biological environment.[26]

While the overwhelming majority of research carried on in the United States adheres to conventional Christian/Baconian models of knowledge acquisition, the search for alternative frameworks for inquiry has slowly been picking up momentum over the past few decades, fueled by the growing recognition that some of the crucial problems facing scientists and humanity alike are intrinsically resistant to the reductionist approach. In physics, for example, the search for, and potential significance of, a "final" or "unified" theory of physical reality is a subject of great philosophical contention. Nobel laureate Steven Weinberg writes: "The convergence of explanations down to simpler and simpler principles will eventually come to an end in a final theory,"[27] while another Nobel laureate physicist, Philip Anderson, noted as long ago as 1972 (thus anticipating Havel by twenty-two years) that "the more the elementary-particle physicists tell us about the nature of the fundamental laws, the less relevance they seem to have to the very real problems of the rest of science, much less to those of society."[28] In molecular biology, the multibillion-dollar program to map the human genome is justified by its supporters for its "role in the localization, isolation, identification, and functional study of genes implicated in human disease,"[29] while others argue that for the great majority of noninfectious diseases, reductionist efforts to correlate particular genetic elements with particular diseases serve no practical purpose and in many cases will prove intrinsically futile: "[Ultimate] behavior (complex disease) is not encoded in DNA but rather in the environmentally interactive cellular epigenetic network which includes, but is not limited to, the genome."[30]

In terms of the environmental challenges that now face humanity, the concept of Gaia offers a particularly revealing—and, at the moment, highly speculative—alternative to reductionist science. The Gaia hypothesis, first articulated by the chemist James Lovelock in the early 1970s, proposes an intimate evolutionary linkage between the physical, chemical, and biological components of the earth.[31] In the Gaian perspective, these components are viewed as interdependent and even self-regulating, perhaps in a manner analogous to the relation between the various organ systems of the human body—circulatory, nervous, digestive, etc.—and to the physiologic regulation of such attributes as body temperature and pH. Lovelock explains: "Gaia theory is about the evolution of a tightly coupled system whose constituents are the biota and their material environment, which comprises the atmosphere, the oceans, and the surface rocks. Self-regulation of important properties, such as climate and chemical composition, is seen as a consequence of this evolutionary process. Like living organisms . . . [this process] would be expected to show emergent properties; that is, the whole will be more than the sum of the parts. *This kind of system is notoriously difficult, if not impossible, to explain by cause-and-effect logic.*"[32]

While this hypothesis may or may not prove useful in permitting scientists to achieve a more synthetic understanding of the environment, at the very least it affords additional insight into why reductionist science may itself fail to attain such understanding. The biologist Lynn Margulis, who next to Lovelock is the scientist most associated with the formulation of Gaia, argues that a lack of interdisciplinary synthesis makes it literally impossible for most scientists even to comprehend the hypothesis: "The Gaian viewpoint is not popular because so many scientists, wishing to continue business as usual, are loath to venture outside of their respective disciplines. At least a generation or so may be required before an understanding of the Gaia hypothesis leads to the appropriate research."[33]

Many skeptics, on the other hand, say that there is nothing much to understand. A *Science* magazine profile of Margulis was subtitled "Science's Unruly Earth Mother," thus casting mystical, New Age aspersions on its subject from the outset.[34] Scientists quoted in the article used words like "fantasy," "simple-minded," "a pretty metaphor," and "grandiose"; overall Gaia is often dismissed by its critics not as flawed science or poor science but as nonscience. Most damning from a scientific perspective are the assertions that Gaia is a moral position, not a scientific hypothesis: "Romantic traditions persist in finding practical lessons and moral directives in natural phenomena and, at least by implication, in urging back-to-nature sentiments or in somehow imitating the cosmic process," writes biologist George C. Williams. "The idea that the universe is especially designed to be a suitable abode for life in general and for human life in particular is, of course, an old one. It had to be abandoned in its early forms with the triumph of Copernican astronomy in the Renaissance, but some scholars still find it possible to argue that the Earth, at least, can be regarded as especially suited for human life. Its main modern manifestation is in the Gaia concept of Lovelock and Margulis."[35]

Of particular interest in this critique is the assertion that Copernicus, in recognizing that the earth does not lie in the center of the solar system, somehow cleared the intellectual decks of anthropocentrism and allowed for a more objective, "premoral" foundation for scientific inquiry, yet this is far from the case. The religious motivation of Renaissance scientists has already been mentioned. Johannes Kepler (1571–1630), who devised the mathematics that proved the validity of Copernican astronomy, wrote: "The movements of the heavens are nothing except a certain everlasting polyphony (intelligible, not audible). . . . Hence it is no longer a surprise that man, the ape of his Creator, should finally have discovered the art of singing polyphonically . . . that he might to some extent taste the satisfaction of God the Workman with his own works, in that very sweet sense of delight elicited from this music which imitates God."[36] (In fact, Kepler—one of the great geniuses of Renaissance science—was an adherent and practitioner of that most anthropocentric of all superstitions: astrology.)[37]

Modern reductionist science was born explicitly of the moral obligation to know the polyphony of nature—and to conquer it. But even scientists who are passionately committed to understanding and reversing the ecologic crisis may strongly resist the notion that the character of science itself is in any way connected to the crisis. Paul Ehrlich writes: "[In] the context of Gaia, I find myself taking a reductionist position; the idea that life evolves in a way to make the planet more hospitable for itself collapses for want of a mechanism."[38] Yet the recognition and scientific study of natural phenomena do not depend on mechanism. There is no known mechanism for human consciousness, nor even for gravity, yet science does not deny the existence of either. Descartes, among others, viewed absence of mechanism as proof of the existence of God;[39] complexity scientists see it as evidence of emergence.

My purpose here, however, is not to promote the Gaia hypothesis—which may or may not stand up to critical analysis in the long run—but to suggest that the conduct of science cannot be morally isolated from the use of science in society. More specifically, Christian/Baconian science has served the moral purpose of advancing the human capacity to act *on* nature by adopting the metaphor of nature as machine, as clockwork, comprehensible and exploitable through the study of its component parts. One need not disavow this moral purpose to recognize its potential lack of suitability for comprehending and confronting threats to global ecologic systems—a crisis of humanity acting *in* nature.

Ultimately, therefore, claims that reductionist science and its technological consequences can save us from ecological crisis should be viewed with skepticism. The reason for such skepticism is not ideological but pragmatic. The idea that "repairing the ozone layer" is nothing more than a "macroengineering problem"[40] implies that the ozone layer is an isolated entity that can be manipulated with impunity and predictable consequence. The claim that the U.S. Global Change Research Program can achieve a "predictive understanding of Earth-system behavior on time scales of primary human interest"[41] suggests that reductionist analysis is the appropriate tool for understanding the behavior of complex systems. Such promises may be intrinsically unattainable. While Gaia may or may not prove to be a useful epistemological framework for looking at these problems in new ways, the controversy surrounding it does illustrate how the direction of scientific progress is partly dictated by cultural roots and how the prospects of society are influenced by these roots even when they are deeply buried in history.

Moral Science

The traditional view of the relationship between science and morals was recently reiterated in a letter to the *New Yorker* magazine written by a graduate student in physics: "Scientific inquiry is guided by one thing:

truth. Its means are the search for truth and its ends are the understanding of it. A by-product of this process is technology, and it is in the production of technology rather than in the practice of science that ethical and social issues must be considered."[42] These words are a fairly good summary of the modern view of the moral significance of science and the linear view of technological innovation; that they were written by one of tomorrow's generation of scientists shows that the perspective continues to propagate. This view is flawed on two accounts. First, science and technology have become so interdependent in the past century that it is impossible to meaningfully separate one from the other or to expect significant gain from one without commensurate gain from the other. Second, the search for truth is not independent from social, political, economic, or moral suasion. The question at hand is not whether scientists search for truth but which truths they search for and how the directions of their search emerge from, and influence, culture.

Sylvan Schweber suggests an alternative view, encompassing both of these concerns: "The scientific enterprise is now largely involved in the creation of novelty—in the design of objects that never existed before and in the creation of conceptual frameworks to understand the complexity and novelty that can emerge from the *known* foundations and ontologies. And precisely because we create those objects and representations we must assume moral responsibility for them. . . . I emphasize the act of creation to make it clear that science as a social practice has much in common with other human practices."[43]

The metaphor of science as an endless frontier serves to isolate the pursuit of scientific truth from its cultural roots and its social consequences. It portrays research as the starting point in a process of exploration, preceded by nothing. Utility of scientific knowledge comes later, as value added to the knowledge. But research and utilization are in fact inseparable; indeed, this is the essence of the Christian/Baconian tradition. The moral injunction to exercise dominion over nature motivates—and is satisfied by—the pursuit of reductionist scientific knowledge.

From a policy perspective, the endless frontier metaphor supports a Panglossian optimism that more research and more knowledge is always and by definition a step in the right direction because, frontiers being what they are, no particular direction is distinguishable from any other.

This viewpoint has served industrialized society well not because it is correct but because the cultural and large-scale momentum of science has generally been parallel to, and thus indistinguishable from, the larger-scale momentum of industrial and economic development. A policy problem lurks here: when the consequences of science conflict with the well-being of society, or fail to meet societal expectations, the source of conflict or failure is inevitably traced to society itself, but the solution is often sought in more science: socioeconomics caused ozone depletion, and reductionist science and technology will fix it. On the one hand, then, the true nature of the science-society relation is obscured; on the other, the range of action that society is likely to take in pursuing its goals is artificially constricted. Chapters 7 and 8 are devoted to a more detailed consideration of these two related problems.

7

Pas de Trois: Science, Technology, and the Marketplace

Nothing obscures our social vision as effectively as the economic prejudice.
—Karl Polanyi, *The Great Transformation*

THE PRODUCTS OF THE SCIENTIFIC IMAGINATION are responsible for
shaping, to some considerable degree, the character of modern existence.
The standard litany of discoveries and innovations made over the past
century—plastics, vaccines, transistors, lasers, fiber optics, recombinant
DNA—serves as one measure of societal progress, progress apparently
derived from the astonishing and ever-expanding variety of new products
and processes that have permitted us to conquer diseases, reduce the
burdens of physical labor, increase the ease and speed of transport and
communication, expand the options for leisure-time activities, widen
access to information and data of all types, and create the seemingly
infinite number of conveniences that consumers throughout the industrialized world take for granted. One may envision even greater progress
in the future, measured by advances in such areas as artificial intelligence, biotechnology, high temperature superconductivity, molecular-
scale manufacturing, advanced materials, and hydrogen power.

The very idea of progress in science and technology implies a sort of
directionality that seems, when viewed retrospectively, to be logical, linear, and incremental. From semaphore to telegraph to telephone to computer network; from stethoscope to X ray to electroencephalograph to
magnetic resonance imagery; from wood to coal to oil to natural gas to
photovoltaic cell—these are all apparently unidirectional trends: things
become faster, stronger, more precise, more efficient, more concentrated,

117

more comprehensive. Such sequences can be interpreted as measures of progress in society; they imply a path of change defined by the resolution of a particular succession of related problems. Thus it seems that humankind is moving forward as well, borne upon the shoulders of the scientists and technologists who make such progress possible.

Such a perspective suggests that the relationship between progress in the laboratory and progress in society is largely unmediated—that the beneficial impacts of science and technology on society derive from the intrinsic attributes of new products and processes. This is the familiar rationale behind the R&D system: that scientific and technological progress is necessary to solve a wide range of societal problems, that in solving such problems the cause of human progress will be advanced, and that human progress can therefore be facilitated by generous government support for research and development.

Of course new products and processes are not automatically introduced into or spontaneously assimilated by society. In most industrialized nations, a diversity of paths from laboratory to society are created by the mechanism of the marketplace. This mechanism suggests a different perspective on the relationship between science, technology, and societal progress. Consider a hypothetical biotechnology company that pioneers a new technique for identifying people who are genetically predisposed to develop a certain type of cancer. Production and application of this new technology generate revenues, attract investment, create jobs that permit more consumption of other goods and services—all of which helps to stimulate more competition, productivity, innovation, and job creation. The result is economic growth. While this scenario presupposes a demand for the product that ultimately emerges from the innovation process—in this case, a new health-care technology—it does not presuppose any particular characteristics of the product. Video games and assault weapons would be equally suitable examples. Furthermore, continued technological innovation must allow all such products to be manufactured at lower cost over time, or it must permit continued improvement of the original products or their replacement by new products, in order to boost demand, consumption, and profits.

In 1957, the economist Robert M. Solow published a paper showing that technological innovation acted as the driving force behind rising productivity and economic growth.[1] In 1987, Solow won the Nobel Prize for this work, by which time his conclusion had become dogma, transcending even partisan politics.[2] A budget document released by the Bush Administration in 1992 stated: "It is now widely recognized that a key to enhancing long-term economic growth in America is improving productivity. . . . The Bush Administration has proposed, over the past three fiscal years, a pattern of investment in areas of research and development that will help to boost productivity and improve economic performance. This budget continues the pattern of aggressive investment in both basic and applied R&D."[3] Less than a year later, one of the first public documents released by the new Clinton Administration sent a similar message: "Technology is the engine of economic growth. In the United States, technological advance has been responsible for as much as two-thirds of the productivity growth since the Depression. . . . [Manufacturers] depend on the continuous generation of new technological innovations and the rapid transformation of these innovations into commercial products the world wants to buy."[4]

These economic perspectives put a different slant on the idea of societal progress and what such progress owes to science and technology. Rather than viewing progress as deriving from the direct application of science and technology to societal problems, the economic frame of reference interprets progress as a function of the enhanced ability of individuals to pursue options, opportunities, and desires. This ability is encapsulated by the term "standard of living," an elusive but widely used concept that commonly refers to the purchasing power available to an individual, quantified by such measures as per capita gross national product.[5] In market economies, rising standards of living are therefore interpreted by economists, policy makers, and voters alike as a concrete indication of societal progress: more people have more financial resources to pursue a wider range of choices in their lives, to meet a wider range of needs, purchase a wider range of goods, indulge in a wider range of leisure-time activities.

In the modern free-market state, science and technology are thus conventionally viewed as offering two pathways to societal progress—one by direct application to societal problems, and the other by catalysis of economic growth and rising standards of living: "Through scientific discovery and technological innovation, we enlist the forces of the natural world to solve many of the uniquely human problems we face—feeding and providing energy to a growing population, improving human health, taking responsibility for protecting the environment. . . . Technology— the engine of economic growth—creates jobs, builds new industries, and improves our standard of living. Science fuels technology's engine."[6]

Such notions of progress are incomplete, of course. Civic ideals such as justice, freedom (in its many forms), and community, as well as ideals of individual welfare such as emotional, spiritual, and intellectual satisfaction and fulfillment, may be conceptually slippery, but they are nevertheless crucial elements of the quality of life. These elements are not well captured by anecdotal descriptions of revolutionary technological innovations or quantitative measures of standard of living. Moreover, a view of the relationship between quality of life and progress in science and technology is particularly difficult to bring into focus because the two distinctive roles that such progress plays in society—catalyzing economic growth and directly solving human problems—are not necessarily complementary.

For example, in the 1930s, a middle-class family might have owned a radio, a Victrola, and a telephone; today it might have home entertainment systems, cellular phones, and personal computers. Where does the progress lie? The fidelity of sound reproduction has greatly increased, but we cannot know whether the analog Louis Armstrong in mono at 78 rpm gave less pleasure to listeners in 1938 than the digital Wynton Marsalis on the CD player does today. We do know, however, that more people buy a wider variety of sound reproduction products that are manufactured more efficiently than ever before, that manufacturers add progressively more value to the cost of raw materials, that they employ progressively more people and pay them progressively better wages. In other words, more people in the 1990s can stock their homes with state-of-the-art consumer products than could do so in the 1930s because the size of the economy has grown and standards of living have risen. Is the progress of society between the 1930s and the 1990s more appropriately measured by the increased technological sophistication and diversity of consumer products or by the increased number of people who are able to purchase them?

In the case of medical care, access to sophisticated diagnostic and treatment technologies is certainly less important for physical and mental well-being than having the financial wherewithal to live in sanitary, uncrowded conditions, maintain a healthful diet, escape urban violence, and pursue an occupation that is not physically or emotionally deleterious. In fact, new medical technologies need not even create direct societal benefit in order to create economic benefit. Consider the accelerating rush to develop advanced technologies that can identify the genetic origins of a variety of noninfectious diseases such as cancers, heart conditions, and mental disorders. These technologies promise—and are already beginning to deliver—profits for the health care industry and generous federal funding for biomedical research.[7] But their capacity to improve public health is far from proven. Most noninfectious diseases are not caused simply by a defect at a single genetic location but in fact reflect complex and poorly understood interactions between multiple genetic elements and the outside environment. This complexity has not subdued the passion of researchers, technologists, investors, and entrepreneurs striving to develop new genetic tests, nor has it undermined public enthusiasm and demand for new diagnostic and predictive technologies. The result may well be the flowering of an economically vibrant sector of the biomedical industry, one that is capable of augmenting the average standard of living yet may make little ascertainable direct contribution to societal welfare.[8] At the same time, genetic testing technologies could plausibly contribute to a range of consequences that adversely affect quality of life, including rising medical costs, declining availability of medical and other types of insurance, and even the creation of a potentially new "scientific" justification for eugenic social policies.*

A high standard of living—more purchasing power—affords one the flexibility to pursue one's personal aspirations more freely. To the extent that personal aspirations include the use and consumption of new technologies, then the existence of the aspirations presupposes the existence of the technologies: we do not desire products that do not exist. Indeed, at any given time the connection between the current state of techno-

---

*The latter possibility is hardly unrealistic, judging by the attention lavished by the public, press, and politicians on such recent eugenic approaches to social policy as Richard J. Herrnstein and Charles Murray's *The Bell Curve* (New York: Free Press, 1994).

logical know-how and one's perception of an acceptable standard of living is largely arbitrary: people in the 1950s, no matter how wealthy they might have been, could not bake their potatoes in a microwave oven. While the evolving products of science and technology may define an ever-advancing horizon of material aspirations, at any given moment in time these aspirations are fixed by whatever happens to be both available and desirable. The capacity to fulfill *current* aspirations gives standard of living its meaning, but the material products that are associated with, and in some ways define, a certain standard of living are obviously very different today than they were in the past. In 1965, a middle-class American family could not imagine life with a personal computer; the absence of the technology was not a source of deprivation. Today, a socioeconomically equivalent family may be unable to imagine life *without* a personal computer; its absence is felt as a deprivation. Because the horizon of material aspirations is constantly moving—fueled by the products of scientific and technological advance—we may feel as though we are better off today than yesterday not merely because the economy has expanded and our standard of living has risen but also because we possess things today that yesterday we could not even imagine.

Indeed, a belief in the intrinsic positive value of technological innovation is essential to the health of the market economy. As one analysis of innovation and industrialization explained: "The long growth in scientific and technological knowledge could not have been transformed into continuing economic growth had Western society not enjoyed a social consensus that favored the everyday use of the products of innovation."[9] This consensus means that daily life must be continually modified by the assimilation of new processes and products, as indeed it has been throughout this century. Because it is the job of the marketplace to foster economic growth, the principal filters on scientific and technological progress are consumer demand and potential profitability. Other filters— curing a certain disease, generating power more efficiently, producing tasty vegetables in winter—are of the second order; technologies may pass through these second-order filters but still fail to pass through the filters of profitability and desirability. Conversely, the creation of adverse effects, such as elimination of jobs, generation of pollution, erosion of personal privacy, or facilitation of violence, need not prevent a given technology from passing into society so long as there is a demand and a potential for profit—and government regulations do not prohibit its use.

Science and technology may therefore make a crucial contribution to economic growth and standard of living without necessarily making a net positive noneconomic contribution to the quality of human life or the welfare of society. I do not argue that there is no such positive contribution; but I do suggest that the marketplace does not provide a mechanism to ensure that this contribution will occur. Rather, the generalized societal benefit created by economic growth may obscure the negative impacts of science and technology on quality of life. While it is likely that a greater percentage of people living in the industrialized world today is free from abject poverty than was ever the case during the past several thousand years, we cannot know whether this group enjoys a higher quality of life than similarly fortunate people enjoyed in the past, or what the direct contribution of scientific and technological progress has been to the emotional, intellectual, and spiritual fulfillment of the average person. Speculating upon historical levels of personal satisfaction or happiness is probably fruitless,[10] but the record of literature suggests that the heights of joy and the nadir of despair have been fairly constant over time and that most people float somewhere in between the extremes. If life is in any sense "better" in the present than it was in the past—if levels of societal welfare are higher—this may not be a reflection of the direct contribution of telephones and computers to personal happiness or satisfaction, but of their contribution to an economic system that helps to shield us from elemental want while affording us a wider range of options in pursuing our aspirations. The fact that a modern urbanite would be miserable living in a cave, hunting big game, and wearing animal skins by no means implies that a well-fed paleolithic hunter was miserable too (at least any more so than today's average city dweller).[11]

The Sound of Invisible Hands Clapping

The underlying relationship between economic growth, scientific and technological progress, and societal well-being is perhaps most clearly illustrated by the role of government in supporting basic research—research that is not obviously connected to particular applications. The ideologies of basic research and market economics in fact hold much in common. During the 1980s, the supply-side, anti-interventionist economic-policy makers of the Reagan Administration looked with favor upon government sponsorship of basic research even as they sought to reduce spending for a broad range of domestic programs (including civilian applied research). The policy rationale behind basic research, founded on a belief in the gradual, serendipitous, and unpredictable diffusion of scientific knowledge into the marketplace, was philosophically compatible with "trickle-down" economics. Government support of the creation of knowledge for its own sake was ideologically acceptable because such knowledge was supposed to translate into more generic potential for innovation, more opportunities for new products and higher productivity, and more generation of wealth without requiring lawmakers or bureaucrats to make choices about specific directions of innovation—choices that could instead be made in the marketplace. The Republican Congress that came to power in 1994 is pursuing a similar approach—slashing civilian applied research and development budgets but maintaining a commitment to basic research.

The raison d'être of both the basic research system and capitalism is the pursuit of growth: growth of knowledge and insight in the one case, and of productivity and wealth in the other. And the key to growth in each case is the self-interested motivations of the individual—of individual scientists pursuing their curiosity and individual consumers maximizing their utility. The cumulative effect of all this selfish action is progress for all. But the analogy goes deeper, in that the rhetoric of both basic research and the free market is rooted in an efficiency ethic that gives primacy to magnitude of growth while viewing direction of growth as intrinsically unpredictable and thus outside the domain of government control. From this perspective it is the job of the government to encourage growth of the knowledge base and of the economy but not to try to influence the character of this growth in any way. "[The] pursuit of science by independent self-coordinated initiatives assures the most efficient possible organization of scientific progress," argued Michael Polanyi in a seminal 1962 article on the philosophy of basic research. "Such self-coordination . . . leads to a joint result which is unpremeditated by any of those who bring it about. Their coordination is guided as by 'an invisible hand.' . . . [Any] attempt at guiding scientific research toward a purpose other than its own is an attempt to deflect it from the advancement of science."[12] These considerations are explicitly analogous to the position of free-market absolutists who view any government interference in the marketplace as ill conceived and counter to the purpose of creating new wealth.

Economists have attempted to quantify the generalized contribution of basic research to economic growth. Such efforts have not yielded precise results—is it really possible to calculate the long-term value of the discovery of the transistor effect or penicillin?*—but few would disagree that, in particular cases, economic payoffs can be very high indeed. General economic arguments are less politically compelling than specific success stories, however, and the language typically used to promote and justify federal expenditures on basic research tends to be concrete, focusing on examples of how such research leads to the resolution of particular problems or the creation of specific avenues of technological innovation: "Rosenberg's research on the potential effects of electric fields on cell division led to the discovery of an important cancer drug; Kendall's work on the hormones of the adrenal gland led to an anti-inflammatory substance; Carothers' work on giant molecules led to the invention of Nylon; Bloch and Purcell's fundamental work on the absorption of radio frequency by atomic nuclei in a magnetic field led to MRI [magnetic resonance imaging]. . . . [Research] designed to answer problems posed by nature, or questions based on scientific hypotheses, may often lead to important technological progress."[13]

These sorts of retrospective, anecdotal portraits—restatements of the myth of unfettered research—introduce a significant distortion into the effort to understand how the research and development enterprise creates most of its value for society. By singling out the practical, positive consequences of particular scientific discoveries, basic-research advocates create the illusion that there is a connection between the serendipitous course of basic research and the specific problems that society most needs or wants to have solved—as if consumers were demanding that the elec-

---

*Nor, for that matter, would it be feasible to calculate the long-term costs resulting from the discovery of the law of special relativity and the consequent development of nuclear weapons.

tronics industry invent transistors so that their radios and TVs could warm up faster. But if progress in basic research is truly unpredictable, then there can be no way to anticipate what particular problem a given scientific discovery might ultimately address. Conversely, there will be no reason to think that any particular problem—say the cure for cancer—will have a higher likelihood of being solved than any other problem—the invention of a new type of weapon, for example. From an economic perspective, this assertion of unpredictability is not problematical because there are no priorities in the marketplace save the creation of economic growth. Faith in the serendipitous progress of basic research is rewarded because of the nature of economic markets—it doesn't matter what gets created or discovered as long as it leads to something that someone wants to buy. In retrospect, this looks to us like progress.

That there are both positive and negative dimensions to scientific and technological progress is obvious. Attempts to balance them may be futile, but an excessive faith in the promise of benefits may dull the sensitivity of society as a whole to the potential for negative consequences. The experience of television—once envisioned as a revolutionary cultural and educational medium, now reviled as the opiate of the masses and the force that transformed democratic dialogue into thirty-second sound bites—does not seem to have dampened anyone's ardor for the coming nationwide high-speed computer network—the so-called information superhighway that will revolutionize data transmission and communication in the United States and create huge new economic opportunities for an array of businesses ranging from phone companies to publishers. As science writer James Gleick explains it: "Once the new highway is in place consumers will be able to send or receive vast amounts of information with great speed—transmitting entire, up-to-the-minute electronic encyclopedias, receiving messages worldwide, seeing current movies as well as the full range of cable and network TV, sending out color images, hearing the sounds of La Scala from Milan."[14] Alternatively, the commercial imperatives and manipulative capabilities of the superhighway may overwhelm its potential cultural and political benefits, as they did for television, and saturate the world with dreck. In any case, there will be benefits, there will be negative consequences, and above all, there will be economic growth.[15] An illuminating example of such a tradeoff was reported in a

New York Times article about Japan's relative backwardness in establishing its own computer networks. One of the explanations offered for the reluctance of Japan to embrace network technology was cultural— that the Japanese "value face-to-face communications [above indirect ones]."[16] What are the costs of failing to overcome this barrier to innovation? "Japan is now waking up to the fact that it is far behind the United States [in computer networking], at a great risk to its economy." If this risk is to be averted, face-to-face communications may have to go.

Whereas the collective action of large numbers of people seeking to maximize their utility in the marketplace leads to economic benefit for all, there is no invisible hand ministering to the wise application of scientific and technological progress. Individual preferences combined with the potential for profit can result in the assimilation of products and processes that do not increase, and may even undermine, societal welfare. Individuals may rationally decide to seek the most advanced medical technologies because they are sick, to buy gas-guzzling cars when fuel prices are low, to log onto the information superhighway because everyone else is doing it. Millions of individuals making such choices—solving individual problems—may lead not only to economic growth but to an increasingly expensive and decreasingly equitable medical system, to increased emissions of greenhouse gases and decreased energy independence, to a decline in face-to-face communication that undermines social comity. Economists often label such impacts "externalities," but another word for them is "reality." Indeed, as commercially successful technologies become inextricably woven into the fabric of society, their negative effects become integral to this fabric as well; often, these threads cannot be removed or even avoided without creating unthinkable disruption. Industrialized society will not, for example, abandon automobiles and electricity in order to reduce the emission of greenhouse gases.

The societal implications of scientific and technological progress are further complicated by the never-ending quest of human beings to satisfy their desire for material goods, wealth, and status. If people weren't constantly eager to trade in last year's model for this year's; if they weren't demanding the most advanced medical treatments, the fastest computers, the smallest videocams; if they weren't seeking to increase their incomes and expand their businesses; if they did not, on the whole, perceive a strong positive correlation between standard of living and quality of life, then economic growth would not occur. Science and technology are thus faced with an economic task that is inherently Sisyphean: to nourish the human need to consume. From the perspective of the consumer, there is no such thing as progress; there is only the infinitely ascending ladder of material and social aspirations. Because everyone cannot be on top, the desire for upward mobility—broadly defined—cannot be stanched. There can be no end to growth. In the industrialized nations, where population has stabilized and 85 percent of the world's wealth resides, private consumption continues to rise exponentially, at a rate of about 3 percent a year.[17]

The implications of continuous, technology-fueled economic growth for the future of humanity, however, are by no means clear. In particular, the question of whether the planet can sustain another century or so of the resource consumption and waste generation that accompany such growth, without also undergoing profound and perhaps irreversible environmental change, is not resolved, but most informed observers acknowledge that such change is at least a reasonable possibility.[18] The optimistic scenario goes something like this: progress in science and technology will lead to increases in the efficiency of energy and material use, production of food, and productivity of manufacturing. This progress will allow the world economy to keep growing without seriously compromising the integrity of the earth's environment, thus permitting a more or less continual worldwide increase in standard of living.[19] The pessimistic scenario does not assume that science and technology will have the time or the capability to save humanity from itself and from the insatiable essence of market economics; it anticipates that present trajectories of development could lead to accelerating environmental and economic crises and mounting political and social chaos, the results of uncontrolled, exponentially increasing resource exploitation, pollution, and population.[20]

The widespread and deeply held belief that scientific and technological advance is directly linked to societal progress creates a strong policy bias in favor of the optimistic scenario. But if this progress is attributable less to specific and direct contributions of science and technology to quality of life than to economic growth and the consumption patterns of market economies, such optimism may not be entirely warranted. The optimistic scenario is further reinforced in modern society because scientific and technological progress are often promoted as necessary to counter the negative impacts of existing technologies—oil can save us from coal; nuclear fission can save us from oil; nuclear fusion can save us from fission. In this manner, the history of technology can be viewed as a series of advances that have allowed humanity progressively to escape the constraints of its environment—to grow more food, extract more minerals, synthesize artificial substitutes for natural resources, produce more energy. This sort of progress is driven not just by technical ingenuity, however, but by resource scarcity, consumer preference, and government action as well. In the absence of economic incentives, technology may not evolve in necessary directions. Indeed, persistently low oil prices in the United States have demonstrated that there is no reason to expect the marketplace to respond spontaneously to the rising curve of atmospheric carbon dioxide. Furthermore, technology can reasonably be viewed as the mechanism by which the constraints of the environment are created in the first place: Without the internal combustion engine there would be no need to find energy alternatives for oil. The question is not whether the world is better off with the internal combustion engine than it was without it; the question is whether this technological house of cards can be sustained indefinitely.

The objection may be raised that the economic effects of science and technology cannot be separated from the problem-solving effects—that more people are better off today not just because they are prospering financially but because scientific and technological progress has boosted agricultural and industrial productivity to meet the needs of a growing middle class for food, transport, and energy. If the Malthusian dilemma— population growth outstripping resource availability—has thus far been avoided, it is certainly because progress in research and development, in combination with the operations of the marketplace, has resulted in more efficient exploitation, distribution, and utilization of natural resources. Without internal combustion engines, plant hybridization, and electricity, among many other (and generally lesser) advances, not only would economies have been unable to grow but the hardship of enormous numbers of human lives could never have been eased. This is especially true in the workplace, where the physical strength of an individual human has been rendered irrelevant to most types of productivity.

But those advances that have contributed to meeting the elemental needs of increasingly large numbers of people do not comprise the major portion of economic activity in the industrialized world; in fact, a principal indicator of a modernizing economy is its decreasing economic dependence on agriculture, natural resource extraction, and low-technology manufacturing of clothing and other goods and an increasing dependence on high-technology manufacturing with its capacity to augment hugely the value of raw materials.[21] An average of about 30 percent of the economic output of low-income developing countries comes from agriculture; in the industrialized world, this value is about 3 percent. In Tanzania, 75 percent of all household consumption is devoted to food and clothing; in the United States, the figure is 16 percent.[22]

If we begin to strip away the camouflage provided by the mechanisms of the marketplace, technology may look less like an agent of inevitable progress and more like a loose cannon. The potential for major destructive impacts cannot be discounted. Environmental effects are only one type of impact; of perhaps even greater immediacy is the proliferation of nuclear weapons in the developing world. Other possibilities are more subtle and perhaps less subject to overt societal intervention. For example, scientific progress and technology development may significantly transform the character and viability of democratic institutions.[23] These changes may come from many directions and include the growing influence and importance of technical experts in the democratic decision-making process; the trivializing influence of mass communication technologies on the quality of political debate; the effects of consumerism on democratic values and sense of community; and the threat to privacy created by computerized compilations of personal information such as credit rating, medical history, and consumer preferences. More generally, science and technology may exert an intrinsically undemocratic influence on society because they require people to adapt to change over which they have no control: society must follow where science and technology lead; the relationship is not one of mutual consent, and there is no going back.

Finally, because the creations of the marketplace are preferentially and inevitably (and virtually by definition) a response to the demands of wealth, rather than to poverty or want, the problem-solving capacity of science and technology will preferentially serve those who already have a high standard of living. Research and development priorities in an affluent society may have little bearing on those human needs or goals that cannot be effectively expressed or served through the marketplace. In this sense, the direct positive impacts of R&D activities on quality of life may tend to become more marginal with time, as science and technology increasingly contribute to superfluity and excess rather than fundamental human welfare. The implications of this dynamic are most striking when considered in the context of the developing world.

Nobody's Partner

The affluence of the industrialized nations—on display for all the world to see through television and other mass communication technologies—has created a remarkably universal vision of material well-being. Science and technology are widely understood to be necessary tools for pursuing this affluence: "All major developmental goals . . . depend to a large degree on the ability of countries to absorb and use science and technology."[24]

Two very pronounced but conflicting global development trends have become apparent over the past several decades. The first trend is positive and constitutes a narrowing of the disparity between industrialized nations and the developing world in basic human development indicators such as life expectancy, literacy, nutrition, infant and child mortality, and access to safe drinking water. For example, between 1960 and 1992 the disparity in life expectancy between the developing and industrialized nations fell from twenty-three years to twelve years, due mostly to a decrease in infant and child mortality. Overall, and on average, basic human needs are increasingly being met throughout the world, although severe deprivation in the South has in no way been eliminated, with over 1.3 billion people still living in absolute poverty, a similar number lacking access to safe drinking water, nearly two billion people without adequate sanitation facilities or electricity, and infant mortality rates still five times greater than in the North.[25]

The second trend is negative. Between 1960 and 1991, those nations comprising the richest 20 percent of the world's population increased their share of the global gross national product from 70 percent to 84 percent. This growing concentration of wealth in the industrialized world has been accompanied by increasing disparity between industrialized and developing nations in such areas as per capita mean years of schooling, per capita enrollment in higher education, scientists and engineers per capita, total investment in research and development, availability of computers, and proliferation of communications technologies such as telephones and radios.[26] In 1990, the United States had 545 telephones per thousand population; Brazil had 63, the Philippines 10; Indonesia 6.[27]

Progress is being made precisely in those areas of global development that are *not* greatly dependent on the generation of new scientific knowledge and technological innovation. Advances in agriculture (the Green Revolution) and medicine (especially vaccinations) have played an important role in meeting basic needs, but continued development progress has been less dependent on new science than on the ability of economic, political, and social institutions to deliver such services as clean water, education, and rudimentary health care. Furthermore, if the human development gap between the industrialized and developing world is slowly narrowing, this is occurring in large part because most basic development indicators have natural maxima and the industrialized world has already reached or is approaching many of these maxima—literacy rates and access to clean water cannot exceed 100 percent; average life expectancy probably runs into a natural limit of considerably less than a hundred years.

In contrast, the industrialized nations are rapidly outdistancing the developing world in measures of human development and standard of living that are dependent on continued scientific advance and technological innovation—especially those based on the creation of wealth and the acquisition of material goods. A nation with many scientists and engineers, many telephones, many computers, many universities, and many high-technology companies will generate more ideas, more opportunities, more productivity, and more economic growth than a nation that lacks these assets. This kind of growth perpetuates itself by demanding more technical expertise, more research, and more information and communication technology, which in turn leads to more innovation, greater productivity, and increased economic growth. Few nations that are not already part of this loop find it possible to join in. In general, the world's wealthiest nations have benefited from several decades of conspicuous increases in concentration of global income, wealth, savings, foreign and domestic investment, trade, and bank lending. Meanwhile, income disparities within individual developing nations have typically increased, as those few who have access to the benefits of advanced education and technologies and to foreign capital are able to command greater proportions of their nations' wealth.[28]

Matters are compounded by large-scale trends in the global economic marketplace. The profitability of modern, high technology manufacturing has become progressively less dependent on those strengths that the South can bring to the marketplace—inexpensive labor and abundant raw materials—and more dependent on those attributes that the South lacks—technological sophistication and a highly skilled and educated workforce. Erosion of the South's comparative advantage in labor and natural resources is further exacerbated because the most highly trained scientists and engineers from developing nations often emigrate to the North, where job opportunities are more plentiful and salaries are higher. This "brain drain" robs the developing world of indigenous expertise that is a crucial prerequisite for economic development.[29]

These trends are accelerated still further by the global emphasis on increased privatization of technological innovation. Efforts to strengthen the world patent system reinforce the gap in economic opportunity that exists between the developing and industrialized worlds. Patents offer a strong inducement to creative scientists and engineers, not to mention entrepreneurs and investors, but they freeze out poorer nations that cannot afford access to patent-protected innovations. The negotiation of scientific cooperation agreements between the United States and developing nations has frequently bogged down over the issue of intellectual property rights. The United States has been unwilling to enter into such agreements without formal assurance that marketable scientific and technological knowledge developed under U.S.-funded programs will, in fact, belong to the United States. Similarly, opposition in the United States to international treaties governing biodiversity and commercial exploitation of sea-floor resources has come in part "because these treaties contained provisions that were perceived as attempts by developing nations to misappropriate the benefits of technological investment by developed nations."[30]

Because science-based technological innovation is widely understood to be the dynamo of economic growth, and fostering economic growth is widely viewed as the principal function of the modern industrialized nation, the products of research and development are inevitably treated as proprietary. A widely distributed 1992 report on science and U.S. foreign policy argued that "incentives for invention and innovation, such as patent laws and intellectual property rights, must be extended and protected around the world."[31] Yet on the same page this report stated that the prospects of developing nations "will depend in large part on the evolution and diffusion of technologies." These two goals are not necessarily compatible. To the extent that nations insist upon maximizing the return on their science and technology investments through patents and other mechanisms, developing nations will be at a disadvantage in acquiring new technologies, modernizing their industrial base, and fueling economic expansion. The economic "cycle of concentration," as one United Nations report termed it, can only be exacerbated.[32]

Science does not provide equal benefit for all people; scientific knowledge can be—and is—appropriated for the economic and political gain of specific nations, trading blocs, and political ideologies. Although basic research is in many ways an international activity, with much cooperation between nations and unrestricted availability of many types of data and research results, the beneficiaries of this research are those nations that already have sophisticated scientific and technological infrastructures and can therefore capitalize on new scientific knowledge, whether created at home or abroad. Thus, in the early and mid-twentieth century, it was the United States—not China or Egypt—that became the preeminent industrial power on earth, in part because it could exploit new knowledge emerging from the research laboratories of Europe. Similarly, it is often noted that the Japanese technological juggernaut of the latter part of this century benefited from the results of research conducted in the United States.[33] More to the point, however, is that very few developing nations have either the physical, intellectual, or economic resources necessary to turn new scientific knowledge and technology (whether created in the United States or elsewhere) into economic growth. It is therefore not surprising that few nations have managed to join the ranks of the industrialized world in the past century.[34] The great industrial and economic successes of recent decades—Korea, Taiwan, Hong Kong, and Singapore—comprise about 70 million people, or less than 2 percent of the developing world's total population. Although many other nations are said to be "industrializing," the economic gaps between these nations and those of the already industrialized North continue, with few exceptions, to grow wider. Of more than one hundred developing nations for which data is available, eighty-one show declining real per capita GDP relative to the North over the past several decades.[35]

Although the gradual creation of efficient domestic economies that can exploit science and technology to create economic growth may offer a route to relative well-being for a few countries of the South, this is not a viable near-term option for most nations. About 80 percent of the world's population lives in the developing world, and 95 percent of the world's population growth occurs there as well. Not only must these nations struggle to create a reasonable quality of life for their people (30 percent of whom live in abject poverty, a proportion comparable to that of eighteenth-century Europe),[36] but they must do so in the face of high rates of population growth—much higher, in fact, than those experienced by the North during the industrial revolution—and the consequent rapid acceleration in demand for basic resources and services. Furthermore, the great majority of people in the South still live in rural areas and still depend directly on exploitation of the local natural-resource base for their income and sustenance. This diminishing resource base must therefore support not only all future increases in standard of living but also a population whose size will double over the next thirty years or so. The industrialized world reached its current state of development in fits and starts and over many centuries, with no fixed vision of societal evolution, no absolute limitations on economic growth, relatively small populations, and virtually unlimited access to the huge—and cheap—resource base provided by the rest of the world. In contrast, the South (that is, the rest of the world) confronts a future in which the earth's capacity for supporting human activity may well become an impassible roadblock to development.[37] As one analyst explained it: "The *increase* in the population of the 40 low-income countries [by the year 2050 will] be about equal to the *total* population in the world in 1960. Thus, the equivalent of a life-support infrastructure superior to the one that took thousands of years to develop [in the North] would have to be put in place within 60 years in order to improve the quality of life in those regions."[38]

The question here is not whether science and technology are capable of contributing substantially to this goal but whether the science and technology *system* now operating in the industrialized world is capable of doing so. The answer seems to be "no," at least in the absence of major institutional and political change. Today, the research agenda of the industrialized world is far removed from the immediate objectives of developing nations. Although in a very general sense there is a shared need to foster economic development while conserving resources and preserving environmental quality, the goals of these two worlds—and the specific strategies open to them in pursuing these goals—are entirely different. The science and technology agenda of the United States government, for example, is dominated by the development of military technologies that are, for the most part, of no conceivable benefit to the developing world. Indeed, the worldwide proliferation of modern weaponry peddled by the United States and other industrialized nations continues to add to the woes of the South. The U.S. space program, though more benign than weapons programs, similarly has little connection to the developmental needs of the South.[39] Yet defense and space together constitute almost 70 percent of the federal R&D budget. Health research in the United States is focused on diseases of specific concern to societies with long life spans, especially cancer and heart disease. Energy research is dominated by programs in fossil fuels, nuclear fusion, and nuclear fission and focused on large-scale energy production. Federal information-technology programs emphasize the establishment of the national high-speed computer network. Environmental research is concentrating on large-scale modeling of global climate change. Basic research in the traditional natural sciences nourishes academic institutions and a high technology industrial base that exist only in the wealthiest nations. Some or all of these research efforts may have the potential to contribute to the needs of the developing world, but such contributions will usually be marginal and fortuitous.

If the developing nations were setting research priorities for the world, they might emphasize such problems as: improving the efficiency, productivity, and environmental soundness of subsistence and production farming and low-technology manufacturing; devising energy-efficient "end-use" technologies for basic needs such as cooking, lighting, and transport; creating small-scale, nonpolluting, decentralized energy-supply technologies; preventing tropical diseases, such as malaria and cholera, and developing better diagnostics and treatments for respiratory infections; reducing the consequences of natural disasters such as floods and typhoons; increasing the effectiveness of reforestation. But developing nations, with less than 15 percent of the world's scientists and engineers and less than 5 percent of the world's total research and development funding,[40] do not have the resources to maintain major programs in areas such as these, while industrialized nations lack the economic and political motivation to pursue aggressively such research goals.

Transfer of technology from North to South has proven to be a complex and unpredictable mechanism for redressing these imbalances. Not only are the products of northern R&D often discordant with the needs of the South, but poor nations generally lack the human resources and physical infrastructure necessary to assimilate high technology products and processes successfully. Although one can envision optimistic scenarios by which poor nations "leapfrog" into the world of high technology, there is not much precedent for miracles on a large scale. Forty years of North-to-South "technical assistance" has been rife with failures, especially in terms of building economic capacity.[41] From the perspective of science and technology policy, the disappointing performance of technology transfer reflects the false assumption that the utility of scientific and technological know-how is context independent—that an idea or process or machine that works in America will work in much the same fashion in Burundi or Bangladesh. In the absence of healthy market economies, the direct negative impacts of new exogenous technologies are often readily apparent and often quite surprising. Who could have anticipated, for example, that a variety of Green Revolution agricultural projects would have the indirect effect of *decreasing* the economic and social welfare of women in many areas of Africa, by eliminating incentives and flexibility that had previously allowed them to cultivate surplus crops?[42] Who could have predicted that high technology weapons exported to developing nations—ostensibly for defense against external aggression—would increase the prestige and viability of armies relative to other institutions, thus making them "attractive to ambitious but impecunious young men, so the military diverted talent from business, education, and civilian public administration?"[43] Introduced technologies ranging in scale from wheelchairs to hydroelectric dams have often created as many or more problems than they have solved,[44] while projects initiated by foreign experts and supported by foreign aid expenditures are often abandoned as soon as they are turned over to local control.[45]

As long as the global R&D agenda primarily reflects the economic and geopolitical interests of the industrialized world, the capacity of science and technology to serve the development needs of the South will not greatly improve and may in fact deteriorate. Consider, for example, the problem of vaccinations. Eradication of smallpox and global efforts to vaccinate children against such diseases as polio, measles, whooping cough, and diphtheria can justifiably be portrayed as one of the great successes of development aid and technology transfer. But the research and development necessary to develop vaccinations against these diseases was not motivated by global concern for public health in the developing world; rather, it arose from the fact that these were diseases of the North as well as the South, so that affluent markets created both a political and an economic incentive for vaccine development. An analogous situation exists today with AIDS, where the ongoing search for a vaccine and other potential treatments reflects the course of the disease in the North even though the greatest long-term public health threat lies in the South. But apart from AIDS, most of the serious diseases of the developing world today are not significant public health threats in the North. Thus, a biomedical research agenda aligned with the disease priorities of the industrialized world has moved progressively farther from the needs of the South. According to one estimate, 90 percent of the global burden of disease is borne by developing nations, while just 5 percent of biomedical R&D expenditures worldwide are devoted to the tropical diseases that cause most of this burden.[46] In the South today, the potential market for malaria, tuberculosis, and other much-needed vaccines offers the promise of meager profits because these diseases occur predominantly among poor people in poor nations; in contrast, the industrialized world offers huge opportunities for profit from new drugs and other medical technologies that are largely irrelevant to the health priorities of the developing world.[47] A stark illustration of this problem is provided by the case of invasive pneumococcal disease, a major cause of pneumonia and childhood mortality in the South. The streptococcus bacteria that causes this disease also occurs in the North—but primarily as a cause of earaches. The pharmaceuticals industry, therefore, is supporting research and development aimed at finding a vaccine that can protect Americans from earaches; in this direction lies the potential for profit. Such a vaccine would neither be designed to prevent, nor necessarily be effective against, the pneumonia-causing form of the virus.[48]

While it may well be true that science and technology are crucial to the development prospects of the South, it is equally true that the world's research enterprise primarily serves the needs of the North through its contribution to the growth of technology-intensive economies. Continued concentration of wealth, scientific know-how, and technological advance in the North dictates that the priorities of the global R&D enterprise will continue to diverge from the knowledge and innovation needs of the South. Thus, the intimate connection between the economic marketplace and progress in science and technology may be an inherent obstacle to the reduction of social and economic inequity at the international level. To the extent that such inequity is a catalyst for environmental degradation and military conflict, the current organization of knowledge production may be seen as contributing to these consequences as well.

A twofold problem emerges. First, the marketplace—which is the principal venue through which the products of science and technology pass into society—provides no first-order mechanism for evaluating or assessing the intrinsic impact of a given technology on society, save for its potential contribution to profitability and growth. Thus, there is no a priori reason to expect that the direct net consequences of science and technology on society will be positive. Second, the symbiosis of science, technology, and the marketplace may skew the R&D agenda away from society's most urgent problems and toward the relatively less compelling needs of those who have already achieved a decent standard of living. If these observations are at least partly valid, then science and technology carried out in industrialized nations in the future may have a progressively diminishing capacity to provide a net benefit to humanity as a whole.

Science as a Surrogate for Social Action

Although conflicts remain, we are beginning to comprehend the molecular background of aging, and to glimpse possible avenues for intervention.— Bernard Dixon, "Age-Old Dilemma," *Bio/Technology*

I cryed out as in a Rapture: Happy Nation, where every Child hath at least a Chance for being immortal! Happy People who enjoy so many living Examples of antient Virtue, and have Masters ready to instruct them in the Wisdom of all former Ages! . . .

*[The] Gentleman who had been my Interpreter, said, he was desired by the rest to set me right in a few Mistakes, which I had fallen into through the common Imbecility of human Nature, and upon that Allowance was less answerable for them.—*Jonathan Swift, *Gulliver's Travels*

PROGRESS IN SCIENCE and technology has created the technical capacity to cure human want on a global scale, at least for the moment. Failures in this regard are problems of allocation—of politics and global economics, of culture and conflict—but not of production capacity. How, then, can one now define the moral purpose of *additional* scientific and technological progress in modern society? The research community continues to do so in terms of human welfare—additional contributions to health, to the environment, to overall quality of life. In a world where food production is sufficient to meet the needs of all humanity, but well over a billion people still do not have enough to eat, where less than 20 percent of the world's population possesses 80 percent of global wealth and accounts for about 70 percent of global energy consumption, the justification for more research seems therefore to imply that future progress in science and technology will overcome the limitations that human nature and political institutions place on societal progress: "Science is inevitably tilted toward future gain. . . . Only utopians can believe that the problems of energy consumption will be solved by a U.S. president advocating a lower standard of living, or a Chinese premier saying we should stick to bicycles because the developed countries have already saturated the atmosphere with CO_2. The solutions [to such problems] will have to be scientific, such as biosynthetic approaches to CO_2 fixation, solar power, organisms that biodegrade pollutants, and cleverer uses of water resources and urban transportation."[1]

The argument is an interesting one because it dismisses as utopian any suggestion that political or social action—the conscious, collective decisions of society—can have a role in solving the problems of humanity while maintaining that it is entirely realistic to expect the supposedly unpredictable path of scientific and technological progress to compensate for the unfortunate frailties of human nature. The argument is mechanistic when it comes to research—"Science is inevitably tilted toward future gain"—while implying that human institutions are capable of progress only insofar as such progress derives from science and technology. From a policy perspective, however, such a position contains the seeds of its own contradiction. If humanity is unable or unwilling to make wise use of existing technical knowledge—knowledge already sufficient to free the world from elemental human suffering—is there any reason to believe that new knowledge will succeed where old knowledge has failed?

Sickness Care

In 1993, the Federation of American Societies for Experimental Biology (FASEB), a leading professional organization of biological researchers, issued a brief report reiterating the importance of basic biomedical research and supporting the linear policy model that has been operating in the United States for the past half-century. The words will sound familiar:

> Society reaps substantial benefit from basic research. Technologies derived from basic research have saved millions of lives and billions of dollars in health care costs. According to an estimate by the

National Institutes of Health on the economic benefits of 26 recent advances in the diagnosis and treatment of disease, some $6 billion in medical costs are saved annually by those innovations alone. The significance of these basic research-derived developments, however, transcends the lowering of medical costs: the lives of children as well as adults are saved, and our citizens are spared prolonged illness or permanent disability. Fuller, more productive lives impact positively on the nation's economic and social progress. . . .

The quest for excellence has fueled the nation's progress in bio-medical research, provided the U.S. with the best health care in the world, and added substantially to our economic growth. Excellence is the hallmark of American science and the reason why it has achieved "world class" status.[2]

There is a certain hallucinatory quality to the assertion that biomedical research saves the nation billions of dollars in medical costs each year at the same time that the U.S. medical system is by far the most costly in the world—so costly as to be unaffordable for great numbers of Americans, so costly that government health programs threaten to become a serious, and potentially disastrous, drain on the federal budget. Similarly, it is hard to know what to make of the statement that the United States has "the best health care in the world" in light of the instability and inequity of the health care system as a whole and the unspectacular health statistics for the public at large. Overall, the state of public health in the United States is, at best, about average in comparison to other industrialized nations. What is the value of the best and most expensive health care if it does not translate into the best health?

Many nations that devote relatively few resources to biomedical research and development and a comparatively small proportion of their national wealth to health care have achieved levels of public health that are no worse, and are often measurably better, than those of the United States. The least affluent of the industrialized nations, such as Greece and Portugal, have basic public health indicators that equal or exceed those of the United States in areas such as life span, infant mortality, mortality of children under the age of five, and maternal mortality. In the past thirty years, while the U.S. government has spent on the order of $120 billion on biomedical research and development,[3] public health has failed to improve relative to most other industrialized nations—that is, standard metrics of health in America have improved no more rapidly than in nations that spend very little on research. This relationship applies as well to health care expenditures as a whole. No nation on earth devotes as much of its national wealth to its medical system, in both absolute and relative terms, as does the United States, where the annual per capita expenditure on health care is almost $3,000. In Greece, for example, the figure is about $370 per person; in Portugal, about $500. Japan, Norway, and Sweden, which have among the highest levels of public health in the world, spend about half as much per person on health care as the United States.[4]

One may argue that disappointing public health trends in the United States reflect flaws in the health care delivery system and have little or nothing to do with biomedical research and development. One can blame these flaws on the greed of insurance companies, the unreasonable expectations of patients, the burdensome regulations of government, the virulence of malpractice lawyers. Of course, rising medical care costs do reflect the rising costs of new technologies and treatments (and their indiscriminate use by doctors and hospitals), so there is a real and direct relation between progress in research and development and the complex economics of the medical system as a whole. And none of these complications discourages research advocates from insisting that more biomedical research is the key to the future health of American citizens, nor do they prevent policy makers from generously funding biomedical research and development or from accepting and often amplifying the claims made on behalf of medical science.

To advocates of biomedical research, the relationship between progress in the laboratory and gains in public health is self-evident. A representative of a pharmaceutical company states categorically: "The increase in life expectancy the last 50 years has been attributable to new medicines."[5] The eminent physician Michael DeBakey agrees: "The advances made by medical research are . . . reflected in mortality statistics over the past century. Life expectancy rose from 34 years in 1878 to 47 years in 1900, to 67 years in 1953 to 75.7 years in 1991. Infant mortality has [undergone] a 70 percent reduction. From 1963 to 1987, death rates from all causes have fallen by 29.2%."[6]

The correlation between progress in biomedical research and development and improved public health, especially as measured by life expectancy, offers an appealing model for how research translates into social benefit, but unfortunately this correlation turns out to be more fortuitous than causative. Reliable data on longevity in the industrialized world reaches back to about 1850, and since that time life expectancy has been gradually and inexorably increasing, from slightly more than forty years in the mid-nineteenth century to current levels that reach the mid-to-late seventies for most developed nations. In the United States, this progress was particularly rapid between 1900 and 1950, when mortality rates declined by almost 70 percent, leading to a jump of about twenty years in life expectancy.[7]

The nature and timing of these major improvements in public health display several significant characteristics. The first is that most of the reduction in mortality over the past century or so has been among children and infants. In the early decades of the industrial revolution, less than half of all children survived to adulthood; today, infant and child mortality stands at about 1 percent in the industrialized world.[8] Second, about 80 percent of the fall in overall mortality rates reflects the progressive decline in death from infectious diseases such as tuberculosis and pneumonia.[9] Third, medical treatment had a negligible, and perhaps even negative, effect on the course of most infectious diseases prior to about 1945, by which time mortality rates had already fallen to the low levels that have characterized the latter half of the twentieth century. That is, diseases such as tuberculosis, pneumonia, scarlet fever, diphtheria, measles, and typhoid were in sharp decline before the advent of antibiotics and effective vaccines. In some cases, new treatments led to a further slowing of death rates, but the overall contribution of medical advances to reduction of mortality from infectious diseases has been estimated at between 1 and 3.5 percent.[10] Fourth, the steep rise in total expenditures on health care in the United States, as well as government expenditures on biomedical research and development, did not begin until the mid-1950s, at about the same time as the rapid decline in mortality rates began to taper off.[11] Fifth, once basic medical care is generally available (as it is in all industrialized nations), increased medical services do not, on average, contribute to further advances in public health. The economist Victor

Fuchs observes: "Those who advocate ever more physicians, nurses, hospitals, and the like are either mistaken or have in mind objectives other than the improvement of health of the population."[12]

As deaths from infectious disease declined in the first decades of the twentieth century, the slack was taken up by stroke, cardiovascular disease, and cancer. Incidence of and mortality from stroke have been falling since the 1930s. Mortality rates from cardiovascular disease peaked in the early 1960s and have been declining since that time. New medical knowledge and technologies have unquestionably contributed to these trends, but, as in the case of most infectious diseases, mortality rates began their decline before the advent and proliferation of effective medical interventions. Furthermore, changes in behavior—such as dietary patterns and frequency of exercise—may have been as or more important than the effects of medical treatment in supporting this decline.[13]

In the case of cancer, some specific varieties, such as Hodgkin's disease, have been successfully controlled, but overall mortality rates continue to increase. Most of this increase is attributable to longer life spans of the population as a whole—we have to die of something, after all—but even when the effects of longevity are eliminated, cancer mortality still seems to be climbing. Experts continue to clash over the interpretation of epidemiological data, the relative influence of environmental and hereditary factors in cancer development, and the efficacy of various cancer treatments, but there can be little question that the contribution of science and technology to reducing death rates from cancer has, to date, been marginal at best. Future progress in fighting cancer may come predominantly from advances in biomedical research and development and medical intervention, but this result has yet to be achieved.[14]

Rapid improvement of public health is an attribute of industrialized society that has had little obvious connection with the state of medical science and the delivery of high technology medical services during this century. While there have been remarkable breakthroughs in the prevention and treatment of specific diseases, such as vaccination against polio and the treatment of respiratory infections with antibiotics, these breakthroughs are superimposed on a society whose health has progressively improved because of factors that are largely unrelated to biomedical sciences.

What are these factors? In the latter stages of the industrial revolution, rising standards of living led to broad improvements in nutrition and sanitation that were crucial elements in the decline of infectious disease. At the same time, gradual increases in resistance to various infections by society as a whole contributed to declining mortality. Today, level of education is the strongest and most uniform predictor of good health and long life span, for reasons that are not fully understood. To some extent, education is a surrogate for standard of living, but years of schooling is a factor more closely correlated with good health than is income level. More education may also translate into more knowledge about how to maintain personal health, yet anyone who can read a cigarette package can learn that smoking causes cancer and emphysema. Those who are better educated seem more likely to act on such information, but why this is so remains controversial. Education level is not strongly correlated with amount of medical care sought and received; that is, more highly educated people do not acquire better health by spending more time at the doctor's office, although they may be better able to request and purchase more effective measures when they do seek treatment.[15]

Based on the experience of the United States, the direction of causation between public health and biomedical research appears to be exactly the opposite of that claimed by research advocates: only when high levels of public health, affluence, and education had already been achieved did the nation begin to invest heavily in biomedical research and development and modern, high technology medical care. More research and development and more medical care may make a tangible contribution to the health of many individuals, but their capacity to improve public health levels throughout society is considerably less efficient, equitable, and certain than what can be achieved by expanding educational and economic opportunity.

Indeed, sophisticated medical interventions may in some cases exacerbate socioeconomic disenfranchisement. Between 1984 and 1989, life expectancy in the United States rose by 0.6 years. For African-Americans, however, life expectancy during this period declined by 0.5 years—the first such decline in this century—and the gap between life expectancy for blacks and whites increased from 5.6 to 6.7 years. Factors contributing to rising mortality—cancer, pneumonia, and diabetes among the elderly;

violence among youths; and AIDS—afflicted blacks more than whites. Factors contributing to mortality decline, such as reduced incidence of heart disease and stroke, benefited whites more than blacks. That is, when mortality from a given condition is increasing in society overall, blacks suffer disproportionately; when mortality from a given condition is declining, blacks tend to benefit less than whites. The greatest contribution to the expanding gap between black and white longevity was heart disease. Although mortality from heart disease declined throughout the population, the decline was more than twice as great for white males as for black males and 70 percent greater for white than for black females.[16]

In this context, consider the development of new technologies for diagnosing and treating prostate cancer—a disease predominantly of late-middle-aged and older males—which may, at an annual cost of many billions of dollars, add nominally to the life expectancy of the average American male. It certainly will not do much for the life expectancy of the average urban black male, however, which in some inner cities is about fifty years or approximately equal to that of the typical citizen of Nigeria or Bangladesh. Nor will such progress address the fact that total, age-adjusted death rates from prostate cancer have been gradually increasing over the past several decades, or that cancer death rates as a whole are rising faster among the poor than the affluent and faster among blacks than whites. Not only do blacks experience higher overall occurrences of cancer than whites (by 8 percent) but they also have higher mortality rates from cancer (by 35 percent).[17]

Or consider the problem of infant mortality, where the United States ranks twentieth among industrialized nations. The poor suffer from this problem at much higher rates than the affluent. Because the problem is partly economic, it is also racial: black infants die at the rate of nineteen per thousand, about the same as infants in Chile, Ukraine, Bulgaria, and Trinidad and Tobago, and a rate about two and a half times greater than that for white infants in the United States. While the United States has made modest gains in fighting infant mortality over the past several years, these gains are largely attributable to advances in research and development, which allow for successful—and expensive—treatment of babies with respiratory distress syndrome. Such technological advances do not address the socioeconomic sources of infant mortality—inequity of economic and educational opportunity and insufficient access to basic medical services such as adequate prenatal care and advising. These are the causes of the conspicuous disparity between white and black infant mortality. This disparity has actually increased in recent years, paralleling trends in life expectancy.[18]

The growing divergence between public health indicators for blacks and whites that began in the 1980s reverses a long-term pattern of public health improvement in the United States. For most of the twentieth century, the health gap between rich and poor progessively narrowed as rising standards of living and education levels reduced elemental deprivation and rapidly expanded economic and social options among the less affluent members of society. The reversal of this trend in part reflects society's increasing tendency to treat scientific and technological progress as the chief determinant of public health. Two principal effects arise from this tendency. First, the introduction of expensive new technologies and treatments contributes to the rising cost of health care, and such innovations become preferentially available to those who can afford the most expensive medical interventions and whose levels of health are already high. (And because whites live, on average, nearly seven years longer than blacks, they will disproportionately benefit from a biomedical R&D system aimed predominantly at the diseases of old age.) Second, and more importantly, when public health is viewed as an R&D issue, the socioeconomic roots of public health problems are obscured and overshadowed while the motivation for social or political action is short-circuited.

Why should this be? When a leader of the biomedical community asserts "the crucial importance and considerable contributions of the nation's medical centers of excellence for the improved health and longevity of our society,"[19] or when the editor of *Science* magazine portrays the Human Genome Project as "a great new technology to aid the poor, the infirm, and the underprivileged,"[20] they are offering progress in biomedical research and development as an explicit policy option in the pursuit of improved public health. This option holds both popular appeal and political potency. No one can doubt that "the heart-lung machine, arterial substitutes for diseased arteries, transplantation of vital organs, diagnostic imaging equipment of extraordinary precision . . . computed axial tomography scan, positron emission tomography, and magnetic resonance imaging"[21] have saved lives and will continue to do so. The technological capacity to hold mortality in abeyance carries with it an irresistible emotional and political appeal, but its effect on public health may, on the whole, be less than salutary, not because there is anything wrong with saving lives through advanced medical intervention but because the promise of technology creates an alternative to other types of action that are more effective, efficient, and equitable. For the reduction of infant mortality, comparatively inexpensive prenatal care—a process of education more than medical intervention—is more effective in terms of both cost and medical outcome than are high technology neonatal interventions.[22] But the delivery of prenatal care to poor women is social policy, not science policy; it is politically dangerous, intrinsically controversial. And it fails to deliver corporate profits. It is no anomaly that the United States has among the highest childhood mortality rates in the industrialized world while also boasting the longest life expectancy for people over the age of eighty-five.[23]

The proposal of something as simple as a seat-belt law mobilizes lobbyists and provokes bitter political wrangling; but no one can object to the development of new techniques for reconstructive surgery on automobile crash victims. Voters and politicians alike are inevitably drawn toward effortless, politically risk-free solutions to difficult problems. Americans want more government spending on health care; it is the portion of the federal budget about which they appear to have the fewest qualms and to which they claim to be willing to devote even more of their tax dollars.[24] And herein lies the problem: "If we seek a health care system that does what people want it to do (regardless of whether that preference is expressed in the market or through political processes), we should expect considerable inequality at the margin in costs per life saved. To the extent that we deem this an undesirable outcome, the way to guard against it is to rule out the *possibility* of relatively high-cost interventions. If the intervention is unknown, society may, in some sense, be better off."[25] This is because, in the absence of the intervention—that is, in the absence of new, expensive, high-technology treatments—social and political institutions, as well as individuals, will have to address problems of public health in terms of their origins in society rather than their solution in more research and development.

At heart, the promise of science—of the inevitable tilt toward gain—offers to free politicians from the minefield of political accountability by guaranteeing a brighter future for the voting public. Perhaps no politician has ever made a conscious decision to support research on infant health or prostate cancer rather than initiatives aimed at addressing the socioeconomic conditions underlying the health problems of the nation, but the political system as a whole will tend to favor the facile, allegedly deterministic course of "more research" over the treacherous and uncertain path of social change. The battle over health care reform, which helped bring President Clinton to power in 1992 and was a part of his undoing in 1994, demonstrates the point. Congress was unwilling and unable to take meaningful political action that would address the health care crisis directly, but even as the fiscally conservative Republicans who took over the Congress in 1994 were making deep cuts in many areas of the R&D budget, they were boosting support for biomedical research at the National Institutes of Health and thus substituting science for political action.[26]

The direct beneficiaries of federal research support—scientists, engineers, laboratory administrators, university presidents—will tacitly encourage this tendency whenever possible, not by opposing political avenues of social change but by offering a painless alternative. Needless to say, the health care industry will favor the same course because *its* health depends upon it. Representatives of the pharmaceuticals industry opposed efforts by President Clinton to control the costs of vaccines and other medicines—social action, that is—by arguing that such controls would reduce corporate profits and require cutbacks in industrial R&D, thus perhaps curtailing future benefits for those very members of society that the cost controls were intended to help.[27] Such arguments are deeply cynical because the R&D priorities of the pharmaceuticals industry reflect market demand, not social need, and because in recent years many drug companies have been plunging increasing proportions of their profits into advertising and promotion in preference to research and development.[28]

Whereas almost any political effort to address social and economic inequities in the United States will spawn debate and partisan rancor, the promise of science to create societal benefit often attracts bipartisan support and is viewed as intrinsically nonpolitical. The high level of expertise of scientists and technologists insulates this promise from challenge by the general public. The belief that "more research" translates into cures for social problems is rarely subjected to the same level of debate and scrutiny as are other types of policy assumptions. Indeed, any suggestion to the contrary is invariably greeted with derision and accusations of antiscientism or Luddism. But you do not have to be a Luddite to question the fundamental assumptions about the connection between scientific progress and progress in society as a whole. In the area of public health, these assumptions remain unproven and unconvincing.

The Best Defense

Prior to the early 1970s, a fundamental tenet of energy policy held that rising economic output was always accompanied by—and in fact required—rising energy consumption. This common wisdom was thoroughly discredited in the years after the 1973 and 1979 oil price shocks. When oil prices rose precipitously, per capita energy consumption, as expected, immediately began to decline. Yet the industrialized economies did not stop growing: between 1973 and 1986, while U.S. energy consumption remained more or less constant, the economy expanded by about 30 percent. The United States, along with other industrialized nations, found that it harbored significant resilience in its capacity to meet domestic energy needs. The historical correlation between growing wealth and growing energy consumption was broken not by the discovery of a miraculous new energy technology but by changes in behavior brought about by sharp rises in oil prices. Individuals and institutions reduced activities that wasted energy while adopting existing technologies that allowed for more efficient energy use: the owners of residences and commercial buildings added insulation to their structures; automobile drivers purchased more energy-efficient cars; corporate managers invested in more energy-efficient machinery, processes, and product lines. What changed fundamentally was not the types of technologies available but the economic incentives to take advantage of technologies that were particularly energy efficient, and to behave in ways that reduced energy consumption.[29]

Major gains in energy efficiency and conservation continued in the United States for more than a decade after the first oil crisis, but falling oil prices eventually undermined the economic and political motivations that had established this pattern. Starting in 1986, the nation's total energy consumption began to climb once again, and a strong trend toward decreased energy intensity (energy use per unit of economic activity) tapered off substantially. A significant latent capacity to reduce energy consumption through assimilation of existing technologies thus remains untapped.[30]

If energy production and consumption had no consequences beyond the provision of power to society, and if prices accurately reflected the effects of energy use, then this behavior would be rational and consistent with the long-term public interest. But patterns of energy consumption obviously have a great influence on the environment, on directions of economic development, and on geopolitics. Oil, which accounts for more than 40 percent of America's energy demand, is the most obvious example. Combustion of oil is the largest source of anthropogenic greenhouse gas emissions. Oil resources are neither infinite nor evenly distributed in the world. Oil imports constitute nearly 60 percent of America's trade deficit.

Before the price shock of 1973, U.S. energy policy was an ad hoc outgrowth of two major political goals—the promotion of nuclear energy through research, development, and regulatory functions and the protection of the domestic oil industry through oil import quotas and other market interventions. After the 1973 Arab oil embargo, the price of imported oil nearly tripled, energy policy instantly became a national priority, and dependence on foreign oil was acknowledged as a threat to national security. The government sought to minimize the economic displacements caused by the rising costs of imported oil by controlling domestic oil prices. Corporate Average Fuel Efficiency (CAFE) standards were imposed on the auto industry to help bring gasoline consumption down, while various programs and incentives were introduced to encourage the adoption of energy-efficient technologies by individuals and businesses. Federal support for nuclear energy research and development (fission and fusion) doubled between 1973 and 1979, while a major new program for solar and other renewable energy sources quickly ramped up to more than a billion dollars a year.[31] President Carter's admonition that the energy crisis was the moral equivalent of war translated most conspicuously into huge government subsidies for flawed energy R&D projects, such as the ill-conceived program to create synthetic fuels from coal.[32] Then, when oil prices fell in the mid-1980s, CAFE standards were relaxed and energy research was curtailed. From 1980 to 1990, inflation-adjusted funding for nuclear power research and development declined by 60 percent, although it remained the dominant element of the energy research portfolio. Support for research on conservation and renewable energy fell by about 70 percent and 90 percent, respectively.[33]

In the late 1980s and early 1990s, environmental concern over acid rain and global warming, in addition to renewed instabilities in the Middle East, created new interest in energy policy. In 1991, the Bush Administration announced its National Energy Strategy, claiming: "One of the keys to ensuring future energy security is reducing U.S. oil vulnerability. Technological advancements are one of the best ways to achieve this, and the National Energy Strategy calls for increasing investments in areas with the greatest potential for reducing oil vulnerability."[34] Energy R&D programs that had been slashed during the Reagan Administration were stabilized, and modest future increases were proposed. Even so, the Strategy projected that the United States would be importing more than 40 percent of its oil in the year 2010, a decrease of about 7 percent from 1991 levels but an *increase* of 7 percent from 1983 levels.[35]

When Iraq invaded Kuwait in August 1990, the United States was more dependent on foreign oil than it had been in 1973. U.S. energy policy had failed to capitalize on the most significant lesson of previous energy crises: that a technologically and economically advanced nation could, over intervals of only a few years, make significant gains in energy efficiency by emphasizing conservation and reshuffling the balance between various existing energy technologies—without sacrificing long-term economic expansion. While almost every other industrialized oil-importing nation pursued policies that kept the price of oil high enough to discourage extravagant oil consumption and encourage the use of more energy-efficient technologies (for example, trains rather than automobiles), the U.S. approach concentrated on R&D programs implemented in a crisis atmosphere and aimed at permitting Americans to maintain accustomed patterns of energy use.

Thus, in January 1991, did American energy policy find its true definition in the Persian Gulf War, where cruise missiles and laser-guided smart bombs were revealed as the most important technological developments for preserving America's energy supply. The billions of dollars spent fighting this war and protecting oil supplies in general, as well as the doubling of the defense R&D budget during the 1980s, can now be seen as a crucial, if ad hoc, component of the nation's investment in energy security. As with environmental impacts, these too are costs that do not appear at the gas pump.

The war with Iraq did not stimulate a renewed national commitment to achieving energy independence. Major energy legislation passed in 1992 was a patchwork of research, development, demonstration, procurement, and incentive programs aimed mostly at accelerating the pace of energy technology innovation and assimilation.[36] As with previous initiatives, this law failed to address the central economic and behavioral factor determining energy consumption patterns—that energy prices do not reflect their true costs to society—and thus it too promised to be generally ineffective. As economist Nathan Rosenberg explained: "Energy-efficiency policies are not likely to be successful without a considerable restructuring of economic incentives that will make it more worthwhile for individuals and organizations to pursue the stated goals of government policy. While the forces of the marketplace, alone and unfettered, are unlikely to provide satisfactory solutions to energy problems, such solutions are far less likely to succeed if they require people to behave *contrary* to incentives present in the marketplace. At the very least, individuals should be forced to internalize considerations of the relevant social costs as well as private costs of consuming energy."[37] This, however, is just what politicians are most loath to do. For example, the U.S. Congress has consistently refused to move toward internalization of energy costs by means of any type of carbon tax, although such a tax has long been recognized as perhaps the most efficient and equitable way both to wean the United States from its dependence on foreign oil and to contribute to a reduction in greenhouse gas emissions from oil and coal.[38] While the value of increased energy independence is almost universally acknowledged, the political will to pursue policies that can most substantially contribute to this goal has been all but absent. But because they must do *something*, policy makers take refuge in the promise of science and technology.

The energy crises of the 1970s and the Persian Gulf War of 1991 were consequences of a fundamental ongoing transition in the relationship between global human development and global geological and biological systems. Industrialization is confronting the finite availability and uneven distribution of oil reserves, as well as the finite capacity of the environment to absorb anthropogenic waste without undergoing unpredictable and uncontrollable change. Developing nations striving to achieve a greater share of global wealth are competing with the industrialized world for the same finite resource base. Free from the direct exploitation of colonialism, most nations of the South may still find themselves at a perpetual and increasing disadvantage in this competition due to the continuing concentration of knowledge and technology in the North. The hugely disproportionate casualties suffered by the Iraqis in the Persian Gulf War were simply another expression of this disadvantage.

The history of energy use and technology is one of gradually increasing efficiency and perhaps gradually decreasing environmental impact.[39] Here is reason for optimism and sufficient justification for healthy investments in research and development aimed at long-term technological evolution. But this history does not imply seamless social and political transitions; nor does it account for the fact that industrialized nations consume about nine times more energy on a per capita basis than the developing world.[40] Oil price hikes and the Persian Gulf War were relatively mild bumps on the path of economic growth in the United States, but instability in oil prices has had a profound effect on the development prospects of many nations of the South, helping to fuel the international debt crisis of the 1980s, for example. The industrialized world may not be able to avoid more serious energy dislocations in the future, as developing nations become more assertive in their quest for economic growth and equity of opportunity. Similarly, the North may be unable to hold itself exempt from the consequences of future global environmental change, although it may be better equipped to adapt to such change than poorer nations.

The framework for continued conflict is apparent. For example, between 1971 and 1992, per capita energy consumption in China more than doubled—both a cause and a consequence of that nation's impressive record of economic growth. U.S. per capita energy consumption remained about constant during this period but still exceeds that of China by almost *thirteen times*.[41] The economic prospects of China's 1.2 billion people will depend, in part, on narrowing that gap. China's fleet of passenger cars is growing at more than 40 percent a year; for the United States the annual increase is 2.4 percent.[42] In the face of finite world oil supplies, how will the ambitions of China and the rest of the developing world for economic growth be reconciled with America's desire to maintain its high levels of wealth—dependent as they are on high levels of energy consumption? The obvious and reassuring answer, of course, is more research: "The solutions will have to be scientific."[43] But the scientific solutions, if and when they come, will not necessarily appear in the form of marvelous and benign energy technologies that can permit endless growth of world economies without destabilizing the earth's environmental or geopolitical systems. Instead, they may be the cruise missiles and smart bombs of tomorrow, the innovations that allow nations with advanced scientific and technological capabilities to enforce their national interest and preserve high standards of living.

America's victory over Iraq offers a short-term vision of scientific and technological enforcement of the national interest but also a long-term prognosis of greater vulnerability and more conflict in an increasingly interdependent world. The energy crises of the 1970s—wrenching experiences that undermined national self-confidence—suggest, in contrast, a positive long-term prescription for energy policy by demonstrating that changes in economic behavior and exploitation of existing technologies can yield rapid and significant improvements in energy efficiency and patterns of consumption without sacrificing economic well-being. Indeed, if the gains in energy efficiency made in the years after the oil price hikes had continued after oil prices collapsed in 1985, America's dependence on Persian Gulf oil could have been virtually eliminated by the end of the decade.[44] America chose technological determinism over political action and in doing so chose as well to pay the true costs of oil on the battlefield rather than at the gasoline pump.

Tailoring People to Taste

Scientists, technologists, and economists often bemoan the public's lack of statistical rigor. They may point to the irrationality of someone who smokes or who drives a car without wearing a seat belt, but who then lobbies Congress to ban an insecticide that causes one additional cancer death for every million people while creating millions of dollars in revenues for the chemical and agricultural industries. They note that federal environmental regulations and clean-up programs costing millions, or even billions, of dollars to implement have often delivered marginal benefit to society while imposing an economic burden that drains resources from state and local governments and reduces profitability in the private sector. They advocate, as an obstacle to such irrationality, processes of scientific, quantitative risk assessment and cost-benefit analysis in order to justify government regulation of products or processes and establish rational priorities for federal expenditures on environmental protection.[45] They emphasize the importance of communicating risks effectively to the public.

While it is widely accepted that "scientific" analysis must be a prerequisite for regulation and restriction of a wide array of products and processes ranging from nuclear power to pesticides, it is rarely acknowledged that the process by which society decides to adopt and use a given technology or material is intrinsically nonscientific—it depends upon the economic behavior of individuals or groups of individuals and the decisions they make about consumption and utility. The cost-benefit mentality thus demands a scientific rationale for restricting technologies but accepts—indeed, embraces—nonscientific reasons for adopting them. That is, the process of consuming technologies resides in the domain of personal choice—of freedom—whereas the process of restricting them is a matter for the experts only. The political scientist Langdon Winner has pointed out that the rush toward risk assessment as a way of "scientifically" regulating the impacts of technology on society amounts to a sort of ethical sleight of hand in which subjective issues relating to quality of life and self-determination are suddenly replaced by the promise of a "rational" quantitative balancing of "costs" and "benefits."[46] Thus, for example, former MIT president Paul Gray maintains: "We must make an earnest and sustained effort to educate the public about the risks and benefits of nuclear power in terms that permit quantitative comparison with other energy sources,"[47]—as if all such risks and benefits were measurable and all opposition to nuclear energy flowed from misperceptions of these risks and benefits and had nothing whatsoever to do with less quantifiable concerns such as the proliferation of weapons-grade plutonium, the generation and disposal of nuclear waste, or the societal implications of ever greater dependence on expensive, centralized, technology-intensive sources of commercial power.

Furthermore, many applications of risk assessment and cost-benefit analysis stack the cards entirely in favor of those who have an economic interest in a given technology or process. Once a new product has been successfully introduced into society, it creates a strong economic constituency comprising producers, distributors, and users; under such conditions, ex post facto regulation or restriction of the product is often politically impossible. (Imagine, for example, trying to restrict automobile use based on the fact that more than forty thousand people die each year on America's roads.) Yet a quantitative basis for assessing the hazards posed by many products can often be attained only after such products become widely used in society.[48] For example, technology is currently being developed that will allow parents to select the sex of their baby through the process of "sperm sorting." This process could have profound demographic and social implications of an as yet unknown nature. "Let's not panic about it until there is some evidence,"[49] says one advocate of the process. Nevertheless, if this technology winds up as the center of a billion-dollar industry for sex selection, the political and economic obstacles to regulation will be greatly amplified.

This discussion is obviously not a defense of the inefficiency of government regulatory procedures or of any particular bureaucratic strategy for protecting the environment and the public health. It is not a justification of statistical illiteracy among the populace, nor an argument against balancing costs and benefits as one way to assess the efficacy of policy decisions. But the effort to measure the risks and benefits associated with a particular policy dilemma may naturally focus debate and decision making on "hard" data and quantifiable, internalized costs while overshadowing or devaluing costs—ideological, esthetic, moral, psychologi-

159

cal, or otherwise—that are less tangible, more disseminated, and often unquantifiable. Thus, while it may be true, for example, that the government is devoting relatively too large a proportion of its available resources to toxic waste clean-up and too little to coastal ecosystem protection or reduction of non-point-source water pollution,[50] policy debate over these sorts of choices avoids the more fundamental underlying issue: how did society arrive at the point where it finds itself having to choose between cleaning up toxic waste sites and preserving the ecological health of coastal waterways? Does a determination that toxic waste clean-up is less cost effective than estuary preservation imply that toxic waste sites are a tolerable by-product of industrialization? Technical disputes over priorities, costs, and benefits successfully obscure the process by which such dilemmas have arisen in the first place: a societal perspective that views progress in human affairs largely in terms of progress in science and technology.

In this way, society's most challenging problems—economic inequity, public health, energy supply, environmental quality, even crime—become redefined as questions of costs and benefits—that is, as technical issues that are best resolved by technical experts. Once, no one worried about how industries disposed of their toxic wastes because society was growing and progressing, and faith in science and technology was being rewarded with such useful products as PCBs and DDT. Now that many urban and industrial areas are virtually saturated with toxic wastes, rigorous and objective assessments of costs and benefits tell us that the risks posed by the wastes are often not great enough to justify the costs incurred by cleaning them up. The dilemmas are no longer those of social and political action; they become dilemmas of technological and economic efficacy. The long-term effect of this process of redefinition therefore is to move political decision making away from the realm of democracy and into the world of technocracy.

Lest such a warning sound overstated, it is useful to consider some of the promises now being made on behalf of the ongoing revolution in molecular biology, which is commonly prophesied as the twenty-first century equivalent of the electronics boom that remade the world after World War II. How does this revolution look to the hagiographers of science and technology? Consider this portrayal in *The Economist*:

In the past two decades scientific discoveries have turned biology from being a discipline dedicated to the passive study of life into one that can alter it at will. Biologists today believe that by tinkering with people's genes, the units of heredity, they will eventually be able to eliminate most of the diseases that now plague the world. Tomorrow, such extraordinary ambitions may seem modest, as scientists start to work on improving a person's genetic lot in life. . . .[51]

Human genetic engineering could also enhance or improve "good" traits—for instance an extra copy of the human-growth hormone gene could be added to increase height. . . . [Gene] therapy might extend life-expectancy. And though scientists still do not understand enough about the genetic processes that make humans intelligent or beautiful, it might eventually be possible to tailor people to taste. . . .[52]

Why should people not use gene therapy to improve their lot in life? Taking a growth gene might be compared to having a face-lift or other cosmetic surgery.

What would Hippocrates have done? He would soon have found that, although many new technologies raise tricky medical, ethical and social problems, they can be managed with legislation and with the right regulatory constraints—just as society decides how to deal with such controversial issues as in vitro fertilization. Given this, it is hard to see why anyone should reject the opportunities that new medical technologies are likely to offer. The reward, after all, could be a guaranteed hale and hearty future for all.[53]

The possibilities are boundless, apparently. According to a report in *Science,* "[It] might be possible to identify people who are prone to commit violent acts by screening for [appropriate] gene mutations, and then treat those individuals with either diet or drugs to counteract the mutation's effects."[54] Other areas of biomedical research are similarly portentous. A series of editorials in *Science* outlined a neuromedical vision of the future:

The need for increased emphasis on brain research is not a sure cure for the ills of the world but it is a beginning. A better understanding of the brain can certainly help us solve such disorders as Alzheimer's, manic depressions, schizophrenia, visual impairments, and hearing deficiencies. It may also lead us to the understanding of more vague and ill-defined responses such as aggressiveness, nationalism, bigotry, and sadism. Knowledge of how much of brain function is native and how much is learned becomes useful if education is to produce a more tolerant and peaceful world. A basic instinct that allowed prehistoric humans to distinguish prey from predator may turn into prejudice against foreigners in an urban world. . . .

Some are repelled by the idea that we can use education or medicines to overcome basic instincts. Others are unwilling to accept the idea that some instincts are anachronistic. Many are concerned that research in any brain area that is controversial is likely to be misused. So much misinformation abounds already that a little truth is unlikely to hurt.

Progress in neuroscience today is breathtaking. It needs more funding, more mutual sensitivity between scientists and laymen, and more speed in converting the frontiers of science to the applications of world anxieties.[55]

The brain is an organ of the body. This fact is obvious to scientists and anathema to many lay people. As an organ perfected over years of evolution, the brain performs its functions admirably. As is true of other organs, however, the brain can go wrong if one of its parts is not made correctly. When the brain malfunctions, some nonscientists tend to ascribe the cause to bad parenting, a poor environment, or evil spirits, whereas a scientist tends to ascribe it to a malfunction in the chemistry of the brain. To the lay person the brain may be so lofty in purpose and so complex in structure that the concept that it is "merely chemistry" is unthinkable, and the idea that it cannot be mended by loving care, unacceptable. Neuroscientists, however, know that a brain affected by a mutated gene may be as unfixable with loving care as a watch with a broken mainspring.[56]

[It] is time the world recognized that the brain is an organ like other organs—the kidney, the lung, the heart.[57]

The proud assertion that "the brain is an organ like other organs" is taxonomy masquerading as insight. One can also say that the Constitution is merely a piece of paper and justice merely an abstract concept. Where does such analysis lead? The arguments put forward in the previous quotations advocate an allegedly benign variety of social engineering, write off as hysterical or unjustified any fear that such engineering might be misused by or be inherently threatening to society,* and ignore the fact that the application of advanced technologies to modification of human behavior would be administered by imperfect human beings subject to the same "anachronistic" behavioral traits as those who would be modified.

Intellectual and esthetic abstractions are not the "soft" predecessors of hard scientific knowledge. Does the mystery of the human spirit vaporize when faced with the bald fact that "the brain is an organ like other organs?" The philosophical basis for modern democratic society derives from an array of shared, highly subjective constructs that have nothing whatsoever to do with science (although they place a high value on scientific knowledge). Reverence for science and technology—and skill in their conduct—in no way implies an equivalent reverence for democratic values, as Nazi Germany and the Soviet Union have amply demonstrated.[58] Much, perhaps most, of the human experience neither flows from nor is explicable by science, and the "anachronistic" instincts that have their origins in some now-unknown evolutionary accidents have

---

*This approach is common. Nobel Prize–winning biologist J. Michael Bishop used a similar tactic in responding to concerns that widespread genetic testing could lead to political abuses, including a renaissance of eugenics: "But surely our society can find the wisdom and means by which to deal with that unwelcome and oft-repudiated prospect. It is, in fact, all too easy to underestimate the decency and courage of our fellow citizens." Oddly enough, the paper in which this statement is made devotes itself primarily to an excoriation of those who would question the intrinsic social benefits of scientific and technological progress— those same voices whose "decency and courage" would ultimately be called upon to restrict and regulate the uses of genetic testing. In other words, while Bishop predicts that "society can find the wisdom and means" to regulate technologies in the indefinite future, he expresses no tolerance for a discussion of such regulation in the present. (J. Michael Bishop, "Paradoxical Stress: Science and Society in 1993," John P. McGovern Lecture on Science and Society delivered at the February 1993 annual meeting of Sigma Xi, San Francisco, Calif., p. 13.)

also motivated human beings toward their greatest achievements, includ-ing the scientific investigation of the natural world. Understanding the physiology of the brain, or the make-up of the human genome, gives no insight into the complex interactions of individuals, ideas, and environ-ment that can lead to the emergence of an Enlightenment or a Holocaust. Will the opportunity to improve one's "lot in life" through genetic engi-neering be related to one's economic status? If so, will this further add to preexisting social and economic inequities? Which "basic instincts" shall we choose to "overcome," and who will do the choosing? Is "aggressive-ness" an "anachronistic" trait when it exists in a tennis star or one's lawyer or only when it is found in an unemployed inner-city teenager? Does the fact that the United States has a homicide rate more than twice that of Germany, four times that of New Zealand and Spain, and eight times that of Norway and the Czech Republic mean that American brains are particularly inclined to chemical malfunction? How will research into the physiology of the brain or the genetics of violence help to reduce the correlation between socioeconomic deprivation and antisocial behavior?

The suggestion that genetic engineering or neurochemical manipula-tion of individuals might be applied to the reduction of crime or poverty offers a new approach to society's economic and social dilemmas: rather than seek to redress inequities through the mechanisms of democracy, we can modify humans so that they accept their "lot in life." The biologi-cal components of violent behavior are indeed beginning to receive in-creased attention from researchers. Scientists and policy makers now talk of the "epidemic" of violence in America, and a National Academy of Sciences report identifies "genetic influences," "neurobiologic processes," and "brain dysfunctions" as potential causes of violent behavior and fruitful areas of future investigation.[59] While no responsible researcher denies the importance of socioeconomic factors in violent behavior, growing interest in the biological component of violence is symptomatic of the larger tendency to move social problems into the scientific and technological arena. The prospect and promise of high technology re-habilitation of criminals might have enormous appeal to voters and pol-iticians seeking a solution to difficult problems such as the decline of social fabric in the inner city.

The promise that "gene therapy" will improve one's "lot in life" is scarcely more reassuring. Explaining the social impact of such technologies by comparing them to "a face-lift or other cosmetic surgery" suggests that an affluent society might pursue this sort of technological advance with abandon. But even if everyone were smart, beautiful, and tall, there could still be only one president of the United States, nine Supreme Court justices, a handful of Nobel laureates, and a few hundred professional basketball players. Social and economic stratification would not end, nor would inequity of opportunity.[60] If everyone were tall, the benefits of tallness would be much diminished. The suggestion that biotechnology could provide a "guaranteed hale and hearty future for all" might well prove to be a recipe for discontent and crushed expectations. We might imagine a population of disgruntled, eight-foot-tall octogenarian wards of the state, looking back bitterly on a life in which the miracles of science and technology failed to deliver personal fulfillment. Or perhaps the desire for such fulfillment could be removed by locating its source in the appropriate gene or lobe, and making the necessary modifications.

It might well be the case that most scientists and technologists do not share the type of vision outlined in the quotations above. Many researchers actively opposed the arms race, but this did not stop the United States and the Soviet Union from carrying the world to the brink of disaster on the shoulders of weapons scientists and engineers. In the case of biotechnology, the potential for creating new wealth is immense. Investors have already poured hundreds of millions of dollars into start-up genetic technology companies that are still conducting research and are thus many years away from delivering a profit and generating revenues. This type of positive feedback between science, technology, and the marketplace warms the hearts of politicians and economists and sustains the careers of scientists. At the same time, however, it creates a societal momentum that can quickly become uncontrollable. Risk assessment and cost-benefit analysis may be called upon to provide cogent evidence for the rationality of this path. Consumers may be uncomfortable with the concept of genetically engineered tomatoes—for a while, at least—but will society have the courage and determination to restrict the application of research and development that promises to generate billions of dollars in revenues and make us all beautiful, smart, peaceful, and law-abiding in the process?

Faith in science and technology may undermine faith in political and social action based on subjective precepts, if for no other reason than that it is more politically expedient to believe in utopian determinism than to engage in the hard work and eternal vigilance required to maintain a vigorous democracy. When faith in science is combined with the moral lesson of capitalism—that individual action dictated by self-interest adds up to social benefit for all—the resulting potion may well be politically irresistible. The problem, however, is that society as a cultural entity is not bound together by individuals maximizing their utility and putting their faith in genetic engineering. Rather, to the extent that people see themselves as a part of a community, they do so because they share a framework of overarching civic myths and subjective values—such as the right to freedom of speech, of the press, of religion, or the importance of such virtues as restraint, tolerance, sharing, and compassion. These binding myths and values are not intrinsically supported or advanced—indeed they may be subverted—by the promise of material progress through the accumulation and application of objective knowledge. In this sense, science may inadvertently join forces with its opposite—the antiscientific elements of cultural relativism—to delegitimate the central myths of democracy: the one by deifying objectivity, the other by deifying everyone's individual brand of subjectivity. In both cases, the binding force of shared myth is weakened and the center may no longer hold.

Science, of course, is the art of the soluble. If society commands its researchers to "cure poverty," "end war," "fix the environment," or "make life simpler," it is unlikely to get much in the way of useful results. Conversely, from a societal frame of reference, the problems that science can solve at any given time are not necessarily problems that society actually needs to have solved. But the myths of the endless frontier, in combination with the imperatives of the economic marketplace, eliminate such concerns by equating scientific advance and economic growth with human progress. This equation tells us that any problem that can be solved should be solved—and the more problems that are solved the better off we will be. And so science advocates continue to call for "more science" to address problems of politics and socioeconomics—problems for which, in many cases, "technical" solutions already exist. Political and cultural institutions might find their goals better served by responding to such problems as if scientific and technological progress had come to an end and the only recourse left to humanity was to depend upon itself.

9

Toward a New Mythology

Some men went fishing in the sea with a net, and upon examining what they caught they concluded that there was a minimum size to the fish in the sea.
—Attributed to Sir Arthur Eddington

SINCE THE LATE 1980s, a growing sense of turmoil has pervaded the government-supported R&D community, brought on by changes in the policy environment such as the end of the Cold War, the decreasing global competitiveness of U.S. high-technology manufacturers, the proliferation of expensive, long-term research projects such as the Superconducting Super Collider and the Human Genome, and the increasing competition for federal research funds.[1] Lurking beneath all such issues is the ongoing federal budget crisis and a political climate unfavorable to continued growth of the R&D budget. National politics contributed to the uncertainty, first as the 1992 election of President Clinton inaugurated a growing federal commitment to joint government-industrial research and to development projects focused on civilian technologies ranging from electric cars to computer networks, and then as the Republican Congress that came to power in 1994 opposed such initiatives, on the grounds that they were unseemly interventions in the marketplace, and reverted to the traditional emphasis on basic research.

Such changes and disputes have created considerable unease within the research community. Basic researchers feel threatened by a growing government involvement with applied research and technology development; defense researchers feel threatened by the disappearance of their Cold War mandate; academic researchers feel threatened by increased

169

competition for federal grants and by the specter of congressional micro-management; government scientists feel threatened by the downsizing of federal laboratories. Everyone worries about the prospect of continued budget cuts. One symptom of these competing fears and interests has been a proliferation of lobbyists hired by research universities, states, and private industry to influence R&D funding levels and priorities. Another symptom has been the rise of congressional earmarks for specific research projects.[2] On the whole, this turmoil may well stimulate considerable reorganization within the R&D system. The policy challenge is to emerge from this period of reorganization with a system whose value to society is enhanced regardless of absolute levels of funding, and I will offer some suggestions toward this end. The policy danger is that reorganization will be driven simply by increased politicization of research priorities and will lead to entrenchment of the existing R&D power structure.

As the words of scientists and science-policy makers quoted herein would suggest, the context for reorganization remains tightly circumscribed by the prevailing myths of science and technology policy. Recent efforts by various government and science-advocacy organizations to reconsider science and technology policy in light of post–Cold War realities have tended to be models for bureaucratic change, improved efficiency, and realignment of existing priorities.[3] These efforts have failed to examine the basic assumptions that underlie the R&D system; if this system is entering a time of significant change, we still do not possess any widely accepted alternative framework for debating this change, let alone defining and directing it. The "endless frontier" still looms as the overarching mythology of science—indeed, reports on science policy still explicitly invoke and endorse the myth.[4] But if the arguments presented in prior chapters are even partly valid, the mere reorganization of funding priorities and bureaucratic structure within the federal R&D system will not address the fundamental paradox that lies at the heart of this book: the lack of correspondence between progress in science and technology and progress in society. Indeed, if the origins of this paradox are traceable, even partially or indirectly, to the myths and assumptions that determine how the current R&D system operates, then improvements in the efficiency and productivity of the system could have the effect of amplifying the discord in the relationship between science and social welfare, rather than diminishing it.

Five Policy Suggestions

The proper role of federal science and technology policy is to foster research and development that serve the public interest. The myths of science policy successfully characterize certain fundamental aspects of the research process itself, but they largely fail to capture the relation between this process and the broader social context in which it occurs. Not surprisingly, then, an R&D policy framework based on these myths excels in stimulating scientific progress, but is less successful in creating progress that is compatible with the needs and capabilities of society.

Advocates of science and technology often argue that failure to translate technical advances into societal progress cannot be blamed on science and technology but must instead be the fault of society itself. I have sought to show that such arguments are meaningless: the nature and direction of scientific and technological progress both derives from and helps to create the society that surrounds it. Indeed, considering the huge impact of science and technology on society, the rather small size of the R&D system relative to other institutions of equal social significance, and the dependence of this system on government (that is, public) funding, science and technology policy should be seen as a particularly potent tool for leveraging social action.

Examination of the policy myths helps to explain why linkages between scientific and societal progress are often weak, but it does not in itself imply a particular recipe for strengthening them. Furthermore, the myths themselves are so integral to the worldview of the R&D community that no amount of argument, however compelling and vigorous, can delegitimate them over the short term. Any effective policy approach to strengthening these linkages must therefore be directed at the related problems of softening up this worldview and encouraging the system to evolve gradually in new directions. The suggestions that follow are meant to illustrate, tentatively and generally, how such an approach might take shape. They focus on the creation of a more explicitly permeable boundary between the laboratory and the surrounding world and thus embrace a reality that the myths of science policy have effectively obscured—that the R&D system is an integrated component of the society that surrounds it. They also serve to identify a few pressure points within the system where existing priorities seem especially unbalanced and opportunities for productive change may therefore be particularly ripe.

The suggestions start with the fundamental component of the R&D system—the individual scientist or engineer—and progressively broaden their scope to encompass research disciplines, policy makers, the public at large, and finally the international community. The general idea is to graft mechanisms onto the system that create a stronger motivation for pursuing, and better tools for recognizing and measuring, direct contributions of science to societal goals. This approach does not imply the need for any immediate or radical restructuring of the R&D system—such an expectation would be practically foolish and politically naive. Rather, it is aimed at providing incentives and knowledge necessary to allow the system to reorganize itself over time, preserving existing structures that are effective while testing and incrementally adopting new ones that seem promising. None of these recommendations is entirely original—indeed, some are entirely unoriginal—and in most cases some degree of action is already under consideration either by the government or by the research community itself. Unfortunately, such actions tend to be insular and are often motivated by parochial interests and political trends that may be short-lived. The point here, however, is that the creative integration and mobilization of a rather small number of generally modest policy actions could begin to allow the R&D system to expand the level and depth of its contributions to an increasingly interdependent global society.

1. EXPANDING DIVERSITY

Efforts to increase the diversity of the R&D community—and especially its leadership—should be redoubled. Many scientific societies, academic institutions, and government research agencies are involved in programs designed to attract more women and minorities into research. As a whole, these programs have met with some success, but women and most minority groups are still substantially underrepresented in the R&D community, and this imbalance is particularly great at the leadership level.

Diversity programs are typically justified both as matters of social equity and by the observation that the quality of the R&D community must suffer if it draws its talent predominantly from a single segment of society—the population of white males. But a more diverse research community might also be more receptive to, and in fact could become an important source of, structural change in the R&D system. The research community is mostly made up of people for whom the connection between their lives, their ideals, and their work is integral and recognizable, even if unacknowledged. One's area of professional expertise cannot easily be separated from one's subjective worldview. Weapons scientists were not the backbone of the disarmament movement; fetal tissue researchers would not be expected to participate in antiabortion rallies. Career choices in the sciences vary by gender and by ethnicity. In 1991, 20 percent of all science and engineering Ph.D. degrees conferred on male students were awarded in the social and behavioral sciences; for female students, the proportion was 47 percent; for African-American students, 57 percent; for Hispanic-American students, 44 percent; for Asian-American students, 15 percent.[5]

The unavoidable conclusion is that the science that one performs is in some way related to one's gender and one's cultural, political, and socioeconomic experience, and therefore that the character of the R&D system—including its structure, goals, and output—is related to such influences as well. While the results of particular experiments may not be sensitive to the gender or ethnicity of the researcher, the choice of experiments, of disciplines, of research objectives, clearly lies in the personal realm. Even if one's worldview is left at the laboratory door, it still dictates which door one chooses to enter. A more diverse R&D community may be, if nothing else, less inclined to adopt uncritically the mythologies and priorities that originated in the midcentury physical sciences community and more likely to loosen restraints on the system and create greater opportunity for and receptiveness toward change.

2. INTEGRATING THE HUMAN ELEMENT

An overriding implication of the discord between the progress of science and technology and the progress of society is that the most difficult problems facing humanity today are of a fundamentally different nature from those of a century ago. At that time—and for virtually all of human history before—ensuring a sufficient supply of basic commodities for local and regional populations was the essential challenge of existence. At the end of the twentieth century, the problem of production has been superseded by the challenge of global distribution. Yet while the cultural, social, and economic institutions necessary to foster more equitable distribution often seem unequal to the task, the explosion of production capacity that offers to free the world from want simultaneously threatens the natural habitat that sustains human life, and may, through its rapid augmentation of societal complexity, threaten political and cultural institutions as well.

The transition from problems of production and supply to problems of distribution and equity, from the challenge of augmenting growth to the challenge of directing, responding to, and controlling it, ought to require as well a transition in the types of knowledge that humanity needs if it is to continue to move toward its most important goals. The new variable at the heart of this transition, of course, is humanity itself, and, more specifically, the relationship between humanity and the revolutionary progress of science and technology. After all, one function of science in the industrialized world has been to isolate natural phenomena from the human context, and a crucial goal of technology has been to reduce the role of humans in the production process. But as problems of productivity are resolved, the question of how most effectively and equitably to integrate all this productivity into society becomes increasingly conspicuous. These obstacles are not independent of the products of science and technology, yet neither can they be overcome merely by expanding the breadth of knowledge in the natural sciences and by stoking the machineries of growth.

The reverence for knowledge that stimulates scientific and technological progress has not included a reverence for knowledge about how this progress occurs, how it can be directed in a manner most consistent with social and cultural norms and goals, and how it actually influences society. Overall, the pursuit of scientific understanding of science itself—of the research process and the path by which knowledge is attained and assimilated by society—is supported at no more than nominal levels, largely as an academic discipline that is divorced from both R&D policy making and the conduct of research and development in the laboratory.[6] Willful ignorance of the increasingly convoluted nexus between science, technology, and society seems to be a theme of modern culture.[7] The common assertion that science and technology have "solved" fundamental human problems, but that "politics" or "behavior" has blocked the effective use of these solutions, bespeaks not merely an uncritical perspective on the science-society relationship but a blinkered view of the domain of scientific inquiry.

The social and political environment into which new products of science and technology are introduced may be as important a component of progress as the products themselves. Seemingly useful results of research and development in the natural sciences and engineering will often fail to produce their desired effect in society as a result of what Paul C. Stern of the National Research Council calls "intuitively attractive but mistaken ideas about behavior: That people will accept experts' risk analysis at face value; that firms will accept and fully implement regulations; that consumers will act on relevant information; and that the free market or quasi-market incentives will work in practice as they do in theory."[8]

Few people are experts in the technical arcana of a given research subdiscipline, but most people act as if they are credentialed experts in human behavior. Professional journals such as *Science* and *Nature* often contain heated exchanges between natural scientists that focus on the ethical, political, and societal implications of their research.[9] Such debates often include "intuitively attractive but mistaken ideas about behavior." For example, natural scientists commonly express the view that public opposition to pesticide use, nuclear power, and other innovations arises from misunderstanding and overestimation of risk. The oft-prescribed solution is for scientists to do a better job of explaining scientifically determined levels of risk to the laity.[10] Social science research suggests, however, that political ideology and trust are more fundamental determinants of public risk perception than technical understanding. Scientists calling for improved communication of quantitative risk assessment are therefore advocating a course of action that might well be labeled "irrational" or "ignorant" by experts in the psychology and sociology of risk perception.[11]

Every major federal effort in science and technology should include a component of cooperative, interdisciplinary research that brings natural and social scientists together to focus on the creation of increased levels of compatibility between progress in the laboratory and social needs and goals. This is hardly a novel idea; Chapter 2 included a quotation dating from a 1947 science policy report recommending "that competent social scientists should work hand in hand with natural scientists, so that problems may be solved as they arise, and so that many of them may not arise in the first instance."[12] More recently, in the context of agricultural research, one analyst suggested that, "when physical and biological scientists work together with social scientists, the outcome can be a sort of practical political economy of technology change: an analysis of power and interests, and of who stands to gain and who to lose, which can both inform and change styles and priorities of technical research, influence the way in which programmes are designed and implemented, and improve the chances that poor people will benefit."[13]

But the R&D system as a whole is organized around the intuitively attractive idea that such efforts are unnecessary, that more knowledge and innovation in the natural sciences and technology will automatically enable humans to create more social benefit. Because this assertion seems like common sense, the very idea that it may, in fact, be a fit subject for critical analysis—that humanity's future prospects may bear on such analysis—has never been a serious component of policy debate. Two related policy problems are implied here. The first is that some types of natural science research are inherently flawed by their isolation from the social sciences. Global climate change research is one example:

> If analysts make erroneous assumptions about how people affect the environment, they may err in estimating rates of environmental change and, perhaps more significantly, by underestimating the uncertainty of their analysis. If they make erroneous assumptions about how the environment affects people, they may neglect feedback processes that might be used to mitigate or adapt to global change. If they make erroneous assumptions about how people use information or about what information they will want, they may misdirect their efforts, perhaps producing information no one needs, or producing information in a way that no one can use it.

Analysts have often made erroneous assumptions of all three kinds. . . .

[Social] science can help global change research by improving the inputs to models of the global environment, providing techniques for improving scientific analysis, and soliciting appropriate outputs for decision making and policy analysis. Such benefits have remained potential rather than actual for a long time.[14]

In response to such concerns, a modest effort to study "human dimensions" in climate processes was initiated as part of the U.S. Global Change Research Project; in 1995, about 1.3 percent of the USGCRP budget, or $25 million, was devoted to such studies. The ultimate impact of this effort will depend not just on the absolute magnitude of the investment, however, but as well on the ability of social and natural scientists to work together and create a mutually reinforcing research agenda. This will require not just money but the development of new structures—both administrative and intellectual—for interdisciplinary cooperation.

The second policy problem arising from the failure to integrate research in the social and natural sciences is that certain areas of research and development, although widely recognized as having a potentially revolutionary effect on society, do not include adequate efforts to understand and plan for future social impacts, monitor and evaluate such impacts as they occur, and modify the research agenda in response to anticipated or actual impacts. The Human Genome Project and the information superhighway each illustrate this problem.

The Genome Project raises the specter of eugenics—modification of the human race through selective breeding—while also prompting concerns about the ethical and social consequences of new types of genetic testing and information. The ethical implications of the Human Genome Project have generated considerable discussion and have fostered productive dialogue between biologists and social scientists. However, while 3 percent of the Project's budget is devoted to a program on "Ethical, Legal, and Social Implications," no mechanisms exist by which this program can directly influence the pace, direction, and application of the biological research that make up the Project's core. Furthermore, as the scientific and economic stakes of the Genome Project grow larger,[15] changing or curtailing the research agenda in response to ethical or other noneconomic concerns will grow less politically feasible. The policy lesson here is that the work of social scientists, scholars of law and ethics, and others who can bring insight into the societal implications of genetic mapping should have preceded and guided the full-scale initiation of the Genome Project.

The Genome Project inevitably attracts scrutiny and attention because it carries both the historical baggage of eugenics and the emotional impact of a research program that is supposedly confronting the question "What makes us human?"[16] The information superhighway, in contrast, has elicited minimal concern and consideration from the public, the government, and the social science community despite the probability that its social consequences will be as profound and far reaching as those of the Genome Project.[17] Two areas of concern loom especially large. First, how can the superhighway be designed and implemented so as to facilitate equity in access to information, in order to guard against its becoming a catalyst for further concentration of knowledge, wealth, and power in society? Second, in a society that is already saturated with data and information, how will the huge increase in both the volume and diversity of information transmitted electronically over the superhighway be productively absorbed? What will be the effect of this further overwhelming of the human capacity to filter and assimilate information? Should critical standards for information and data be established, and, in the absence of such standards, will information of high quality or utility be lost in a sea of useless or trivial communication? Ultimately, the superhighway promises greatly to surpass the combined economic and cultural impact of television, radio, and all existing computer networks, yet the federal program to help develop the national high-performance computing network includes no integrated component that investigates social impacts let alone tries to incorporate social effects research into the design of the system.

3. HONEST BROKERS

Politicians are experts at evaluating the interests of their constituents. (Those who are less than expert at this task rarely stay in office for long.) Scientists are experts at creating new knowledge and insight about nature. But neither politicians *nor* scientists are experts in creating linkages between the R&D agenda and societal welfare. The political brilliance of the myth of the endless frontier lies in its circumvention of the need for such expertise by postulating a direct and automatic connection between political wisdom and scientific insight: so long as politicians generously supported the creation of new knowledge about nature, the public interest would be served.

Greater compatibility between the public interest and the goals of research requires progress in three areas. First, the demands and expectations of policy makers must be more fully grounded in reality—politicians must better understand what they can reasonably expect to reap from research so that they can better articulate realistic milestones and objectives in the programs that they create and better evaluate progress toward their stated goals. Second, the R&D community, for its part, must neither promise more than it can deliver as a tactic for obtaining government support nor use its intellectual and professional standing as a pretext for distancing itself from the public agenda, public accountability, and open debate. Third, as already discussed, new knowledge about the social and behavioral implications and applications of scientific and technological progress must be developed and integrated into the policymaking process. These, of course, are ideals that can never be fully achieved, but under the current mythological framework for R&D policy they are not even acknowledged as necessary or useful.

Indeed, both the R&D system and the political system contain explicit disincentives to each of these recommendations. If policy makers lower or change their expectations for science and technology, then they will find it more politically difficult to defer policy action in favor of "more research." If the R&D community moderates its promises, then it will have greater difficulty justifying the budgets upon which it depends. If research programs require a fuller integration of social and natural sciences, then the pecking order in the R&D community must change and guiding assumptions about the relationship between science, technology, and society—assumptions that protect the status quo—will no longer be sacrosanct.

New institutional structures may be necessary to help create and sustain a better harmony between social goals and the research agenda. One possibility would be for the government to support a series of small, independent, satellite policy institutes, each of which would examine existing and prospective linkages between a particular social or policy concern and related research programs (for example, global climate change, energy, public health, information and communications infrastructure). These institutes would act, in effect, as honest brokers, responsible for integrating information coming from the policy realm and the laboratory, identifying areas of conflict and areas of consistency between the policy agenda and the research agenda, evaluating research progress over time in light of articulated milestones and objectives, and recommending how both policy goals and research strategies might be modified to achieve greater compatibility with each other and with the public interest.

Specifics of the organization and operation of such institutes would be drawn, in part, from the experience of existing policy analysis organizations and from successful methods developed in areas such as the policy sciences and technology assessment. But the idea of disinterested, independent intermediaries between the research community and policy makers, organized around particular, long-term societal problems, has yet to be tested.[18] Some trial and error would be necessary. To the greatest extent possible, these institutes should be shielded from the influence of policy makers and researchers, although analysts at the institutes would work directly with both groups. Of course policy makers would not be obliged to make use of the information and recommendations that they received, but at the very least they would have access to a type of analysis that today rarely finds its way into the process of making science policy, the kind of analysis that would help to offset the impact of untested policy myths and assumptions. A small number of institutes could be initiated in areas of particular national priority, as part of a pilot project to judge the feasibility of the concept and to help develop a preliminary model for the structure of additional institutes.

The U.S. Global Change Research Program seems to be particularly fertile ground for such an exercise. Certainly there has been a recognition among numerous policy analysts that the putative political and social goals for this program are largely unrelated to the actual research agenda.[19] The Joint Climate Project to Address Decision Makers' Uncertainties represented one modest effort to illuminate these discrepancies. This project, cooperatively funded by government and private-sector organizations, established a "systematic and iterative dialogue between researchers and decision makers . . . to define what scientific information can be provided over various time frames, in order to address pending policy questions on global climate change."[20] The methods for achieving this objective were straightforward, "and included a series of interviews, workshops, and focus groups [with policy makers and researchers] that resulted in a consensus set of policy-relevant general questions for researchers to address."[21]

As unlikely as it may seem, the current organization of the R&D system includes no effective institutional mechanisms for seeking areas of intersection between what policy makers want to accomplish and what researchers can reasonably achieve. More direct, unmediated communication between scientists and policy makers will not solve this problem because such communication simply leads to the definition of congruent areas of political and professional self-interest, as scientists lobby for more research funding and politicians look for ways to avoid making risky decisions. Disrupting this symbiosis requires the intervention of a third party. The point is that a successful research program may depend less on the amount of science generated than on the degree to which that science is successfully integrated and coordinated with a policy agenda motivated by social goals.

4. INTRODUCING DEMOCRACY

It is axiomatic that a scientifically literate public is crucial to the welfare of modern, industrialized nations: "America's future—its ability to create a truly just society, to sustain its economic vitality, and to remain secure in a world torn by hostilities—depends more than ever on the character and quality of the [science] education that the nation provides for all of its children."[22] But *how*, exactly, does America's future depend on a more scientifically knowledgeable public? Promoting science literacy and science education has become a veritable cottage industry among leaders of the research community, but this promotion is characterized by a singular focus on the one-way transmission of information, ideas, and knowledge from the laboratory and into society: "What the future holds in store for individual human beings, the nation, and the world depends largely on the wisdom with which humans use science and technology. But that, in turn, depends on the character, distribution, and effectiveness of the education that people receive."[23]

The converse position is rarely heard: that the effectiveness of science and technology in a democratic society could depend on the information that the R&D system receives from the public. After all, the important and often controversial policy dilemmas posed by such issues as nuclear energy, toxic waste disposal, global climate change, or biotechnology cannot be resolved by authoritative scientific knowledge; instead, they must involve a balancing of technical considerations with other criteria that are explicitly nonscientific: ethics, esthetics, equity, ideology. Trade-offs must be made in light of inevitable uncertainties. If the public is not directly involved in deciding which trade-offs to make, then the influence of experts, acting within their own subjective frame of reference, can only grow more significant over time, while citizens become increasingly cut off from the forces that shape their lives—even if their understanding of these forces is better than it used to be. At some point, of course, this trend could provoke a true, and truly destructive, antiscience, antitechnology backlash. With the public largely deprived of any significant voice in the management of science and technology policy, and the effects of technical progress on daily life constantly increasing, one key to avoiding such a backlash may be greater public enfranchisement.

At present, most citizens have only two options for involving themselves in decision making about science and technology—the diffuse mechanism of voting and the direct but often unmediated local action that is commonly associated with not-in-my-backyard sentiments. A middle ground that enhances opportunities for public participation, while also providing mechanisms for technical input and open dialogue between scientists and the laity, remains to be defined. Nevertheless, just as society often counts on nonexpert juries to make reasoned decisions about difficult legal problems, so are interested members of the public— working with technical experts and mediators—quite able to acquire the background and information necessary to make reasoned and balanced decisions about complex technical issues.[24] This process does not depend on the prior public understanding of Newton's laws or Darwin's theory of natural selection or any other arbitrary measure of science literacy. It does depend on the creation of avenues by which the public judgment can be brought to bear on important issues of science and technology policy, and on granting the public a stake in the decision-making process. The policy goal is not to substitute "common sense" for technical knowledge but to allow democratic dialogue to play its appropriate role in decision making that is inevitably dominated not by authoritative data but by subjective criteria and social norms. Progress toward this goal is important as a matter of democratic principle and also as a route to breaking down the isolation and thus expanding the worldview of the R&D community.

In 1992, the nonprofit Carnegie Commission on Science, Technology, and Government recommended the creation of a National Forum on Science and Technology Goals. The forum was envisioned as a venue for public participation in the definition of national R&D goals and a mechanism for incorporating public opinion into the science and technology policy-making process.[25] Although the centralized nature of the proposed Forum would have precluded widespread public involvement, and the proposed siting of the Forum at the National Academies of Sciences and Engineering would have created too direct an affiliation with the R&D power structure, the very fact that a mainstream policy organization advocated such a course of action is reason enough for optimism.[26] Alternative approaches might provide for regional or local forums that bring together members of the public, technical experts, and policy makers to discuss policy options surrounding issues of particular immediacy. Such issues could range from the very local—such as the siting of a solid waste disposal facility—to the national—for example, balancing federal R&D expenditures for nuclear fusion and for renewable energy sources. Perhaps such forums could be associated with and administered by the policy institutes discussed in the previous section. An intellectual framework for structuring such mechanisms already exists,[27] and formal institutions for public participation and public-expert dialogue in technology policy have been established in Sweden, Denmark, the Netherlands, and the United Kingdom, where they address problems ranging from genetic engineering to the adoption of new industrial technologies.[28]

5. THE GLOBAL R&D COMMUNITY

The R&D agenda of the United States, focusing as it does predominantly on military technology, space exploration, basic research in the natural sciences, and the medical problems of an affluent, long-lived society, is almost totally disengaged from the problems that face the developing world. From a global perspective, this disengagement represents an abject policy failure; from the philosophical perspective of the Enlightenment, it is a direct contravention of the principle that scientific progress benefits all humanity.

The effort to redress this disengagement simply by exporting knowledge and innovation from the industrialized to the developing world—technical assistance, in other words—is now generally understood to be inherently flawed because it ignores questions of compatibility between northern product and southern consumer. The source of policy failure is often traced to the lack of a developing-world infrastructure adequate to assimilate northern technology: "[The] recipient of the technology must have competence not far removed from that of the provider if the transfer is to be successful. . . . The lesson for developing countries is clear: indigenous capacity in science and technology . . . [is] an essential requirement for embedding technology successfully in an economy."[29]

Alternatively, southern infrastructure can be viewed not as a problem but as a neglected resource: "[Local] farmers, artisans, and traders possess an extensive knowledge of both technical and demand factors of great importance for technology design and development."[30] According to this view, technical assistance has often foundered upon the inability or unwillingness of northern scientists and engineers to utilize this resource: "[The] body of empirical knowledge of the traditional sector has practically no connection with the R&D systems of the modern parts of society."[31]

These diagnoses are complementary views of the same failure. The products of the northern R&D system are often unsuited for assimilation by southern infrastructure, and either or both could be modified to create a greater compatibility. Subsistence farmers do not need electricity supplied by nuclear power plants to run their water pumps, and nuclear engineers are not paid to design small pumps that run off of photovoltaic cells or wind turbines. From this perspective, the policy problem can simply be seen in terms of a need to foster the communication and cooperation necessary to identify, define, and solve specific problems. The dilemma is practical, not ideological: knowledge possessed by scientists and engineers in the North often cannot be translated into ready-made solutions to development problems, while local experience and expertise may be equally insufficient for addressing the tidal wave of development issues now overwhelming the South. Both sides have much to learn, and herein lies a fertile and largely unexploited territory for productive collaboration.

For example, on the Indonesian island of Bali, Green Revolution rice-farming techniques have threatened the local agriculture system by over-taxing irrigation capacity and creating pest infestations. These effects can hardly be blamed on lack of indigenous technical capacity. Development experts forced Balinese farmers to abandon traditional techniques for water management, which were linked to Balinese religious rituals, enforced by religious leaders, and thus ostensibly antiquated. However, these traditional practices represented an empirical solution to water and pest problems that had evolved over many centuries. On Bali, the key to agricultural success appears to lie in the combination of ancient water management techniques and modern, high-yield rice varieties.[32] (Moreover, what appears to be the solution to the Balinese rice problem was recognized through research in the social sciences, although social and behavioral scientists—with the exception of economists—have typically played a minor role in implementation of the technology-assistance paradigm.)

Similar examples abound,[33] and they suggest that the successful application of science and technology to the problems of the South should be viewed not in terms of "assistance" or of building indigenous "competence" but in terms of cooperation and interdependence among groups that possess fundamentally different, but not incompatible, bases of knowledge. As one Indian industrialist explained: "We in India need things that we can repair, not black boxes that have to be replaced. We need to install technology slowly, in stages. And there should be some link between the introduced technology and the local community. The local area should provide some resource, or the process should use some local knowledge."[34]

The momentum of the American R&D system will not easily be changed, and the political obstacles to meaningful U.S.–developing world collaboration are significant. However, it may not be excessively quixotic to visualize an incremental realignment of research priorities over the next several decades. The issue of global environmental change, which is connected to many of the most significant development problems facing the nations of the South—such as energy use, deforestation, and population growth—has considerable political potency in the United States and other industrialized countries. (In the United States alone, it attracted $1.8 billion in R&D funding in 1995.) Foreign "aid" is notoriously unpopular among American voters, but international scientific and technological *cooperation* carries no such stigma, and true cooperation must become the new paradigm for addressing developmental needs in the South. Public support for both environmental protection and scientific research is strong, and global climate change can provide a useful political symbol for linking the prospects of the South to the continued prosperity of the North; if effectively tapped, this symbol could translate into public approval for increased levels of collaborative research and development on global environment and development issues, perhaps paid for by reducing expenditures for ineffective aid or research programs.

Lacking are institutional and intellectual mechanisms for action. Although collaboration between researchers in the United States and the developing world is not uncommon, such collaboration is usually driven by U.S. science priorities and rarely includes sources of knowledge at the local level, be they rice farmers, small business owners, textile weavers, or religious leaders.[35] Much preliminary effort and experimentation will have to focus on the creation of networks and modes of communication that can facilitate productive collaboration between different types of experts. Development bureaucrats in the North, steeped in the mindset of "aid" and "assistance," are unlikely to adapt easily to an ideal of technical cooperation based on equal partnership, or even to accept that such partnership is possible. New programs therefore should probably be independent of traditional aid organizations such as the U.S. Agency for International Development. Institutional support could come instead from small, bilateral or multilateral, autonomous research endowments capitalized by contributions from each participating nation and designed to support research and development on a particular range of high-priority problems. Several endowments of this nature already exist to support collaborative research between scientists in the United States and developing nations, including India and Mexico; this model must be expanded to incorporate local users and experts. All or most of the awarded funds should be provided in the currency of the participating developing nation or nations, and all or most of the research should be carried out there as well, in order to ensure that the research agenda emerges from and remains consistent with local goals.

Yet the idea of collaboration can and should go further still. The technical-assistance paradigm has assumed not merely that the South should follow a development path more or less similar to that of the North but also that the social, cultural, and institutional structure of the South has nothing to offer the North in return. Both assumptions may be incorrect and shortsighted. The technical challenge in the South is to devise and exploit knowledge and tools that can help to create a decent quality of life for its citizens without requiring them to wait around for the levels of economic activity characteristic of the industrialized nations (presuming even that this could be achieved on a global basis). Still, the future welfare of humanity may depend not only on the capacity of the South to raise the standard of living of its people without destroying its natural environment, but as well on the capacity of the nations of the North to maintain a high quality of life while learning to do with less—less consumption of material goods and energy, less consumption-based economic growth, less generation of waste. In other words, both North and South must strive to identify and pursue development paths that decouple, to the greatest extent possible, quality of life from absolute levels of consumption and growth. In this sense, the long-term developmental prospects of the North and South may depend on the degree to which they are willing to embrace this goal together. While the developing world must certainly benefit from scientific and technological expertise that is concentrated in the North, the industrialized world may be able to learn much from people and cultures that have never had the dubious luxury of defining societal welfare in strictly material terms.

CHAPTER NINE

The Search for Ellipses

The recognition by Nicolaus Copernicus in the early sixteenth century that the earth did not lie in the center of the universe, but in fact rotated around the sun along with the other planets, must be counted as one of the great conceptual leaps of science. Judged by the standards of science, however, the Copernican view of the universe was hardly more successful than the centuries-old Ptolemaic system that it replaced. Neither system could yield mathematical descriptions of planetary motion that conformed precisely to actual astronomical observations. Both required unwieldy, ad hoc mathematical and geometric explanations to account for differences between theory and observation.

Copernicus and Ptolemy were shackled to a universe of circles. Circles were the simplest, most elegant geometric form; intrinsically perfect, they were the only conceivable expression of God's hand in the motion of celestial bodies—so had every great astronomer believed since the time of the ancient Greeks, and on this belief did Ptolemy and Copernicus each founder. In the early seventeenth century, the circular shackle—conceptual, spiritual, even moral in its hold over the intellect—was finally broken when the German astronomer Johannes Kepler abandoned the perfection of the circle and discovered that the behavior of the planets would be perfectly described—the conflict between theory and observation resolved—if planetary orbits were elliptical in shape. Thus did Kepler secure Copernicus's reputation (not to mention his own) for posterity.

Copernicus and Kepler were somehow able to view the universe in a new way; to abandon unexamined assumptions that no one else even recognized as assumptions; to transcend not merely convention but the common conception of reality. Of his unwillingness to forsake a universe of circles Kepler wrote: "This mistake showed itself to be all the more baneful in that it had been supported by the authority of all the philosophers, and especially as it was quite acceptable metaphysically."[36] The historian Lynn White suggests a subconscious dimension to Kepler's inspiration: "The first ascertainable oval design in a major European work of art is the paving that Michelangelo designed in 1535 for the remodeling of the Capitoline Piazza in Rome. Michelangelo and his successors during the next fifty years created an atmosphere in which ovoid forms became respectable, until finally Baroque art was dominated by the oval. Kepler's astronomical breakthrough was prepared by the artists who softened up the circle and made variations of the circular form not only artistically but also intellectually acceptable."[37]

A world of circles can only be understood and acted upon in terms of circles. When Ptolemy and Copernicus each found that circular orbits failed to explain the behavior of the planets, they sought to resolve inconsistencies not by investigating other possible orbital forms but by introducing epicycles—additional, smaller-scale, circular planetary motions—into their theories. In doing so, they preserved their worldview but strayed progressively further from the ultimate solution to their problem.

Perhaps the most influential postwar study of the nature of scientific progress is Thomas Kuhn's *The Structure of Scientific Revolutions*. Kuhn outlines the intellectual conditions that typically precede major scientific breakthroughs, wherein a theoretical model—or paradigm—that guides research activities in a particular field is increasingly unable to explain new types of data or answer new types of questions. The Ptolemaic paradigm, for example, could not adequately account for new and more precise observations of planetary motion. But scientists are bound by the intellectual constraints of the reigning paradigm: "Given a particular discrepancy, astronomers were invariably able to eliminate it by making some particular adjustment in Ptolemy's system of compounded circles. But as time went on, a man looking at the net result of the normal research effort of many astronomers could observe that astronomy's complexity was increasing far more rapidly than its accuracy and that a discrepancy corrected in one place was likely to show up in another."[38]

Although the analogy is not exact, science and technology policy today displays the crucial elements of intellectual crisis that Kuhn suggests are precursors to, and necessary stimulants of, a new worldview or paradigm change. In this case, the ruling paradigm is the endless frontier and its affiliate mythologies. Guided by this paradigm, society continues to throw science and technology at its accelerating scientific and technological problems, and it continues to prescribe more growth to correct the consequences of previous growth. Technical recommendations become increasingly ad hoc as the discrepancy between scientific progress and societal welfare widens: we could launch huge aluminum and zinc screens

189

into the stratosphere to adjust for ozone depletion; we could dump iron into the ocean and promote planktonic consumption of CO_2 to adjust for global climate change; we could cut down more tropical forest, irrigate more desert, and apply more pesticides to boost agricultural production and provide for exponential population growth; someday perhaps we could even manipulate a gene to correct antisocial behavior among the economically disenfranchised.[39]

In 1993, a committee from the National Academy of Sciences, made up of spectacularly credentialed men[40] from academia and industry, arrived at the following three goals for American science and technology policy: "The first goal is that the United States should be among the world leaders in all major areas of science. . . . The second goal is that the United States should maintain clear leadership in some major areas of science. . . . [Third, we] have recommended that the federal government adopt the goal of maintaining a leadership position in those technologies that promise to have a major and continuing impact on broad areas of industrial and economic performance."[41] In 1994, the Clinton Administration announced the goals that would underlie its science-policy agenda: "Maintain leadership across the frontiers of scientific knowledge. . . . Enhance connections between fundamental research and national goals. . . . Stimulate partnerships that promote investment in fundamental science and engineering and effective use of physical, human, and financial resources."[42] There is nothing necessarily wrong with any of these goals, just as there was nothing wrong with sixteenth-century astronomers adding another epicycle or two to better explain the observed position of the planet Mars. But they fail even to acknowledge, much less confront, the roots of the intellectual crisis of science and technology policy: a paradigm that promotes the creation of discrepancies at a far faster rate than our capacity to adjust or respond to them.

The way human beings—even scientists, and perhaps especially scientists—view the world is severely bounded by particular interpretations of reality that Kuhn calls paradigms but that, in a different context, could equally well be labeled myths. These interpretations are crucial parts of rational inquiry and daily decision making; without their constraint, people would be immobilized by an infinity of options. Because it is so difficult—impossible, really—for any one person truly to "see" any more than the tiniest slice of reality, people depend on interpretations, paradigms, and mythologies to understand the world. Wrote Walter Lippmann: "[The] real environment is altogether too big, too complex, and too fleeting for direct acquaintance. We are not equipped to deal with so much subtlety, so much variety, so many permutations and combinations. And although we have to act in that environment, we have to reconstruct it on a simpler model before we can manage with it. To traverse the world men must have maps of the world."[43]

The way that we view the world determines the way that we draw our maps; the way we draw our maps will determine the paths we can follow in the pursuit of progress. In a growth-oriented society, we pursue knowledge that fosters growth, and our maps simply point us toward the endless frontier. The economist Kenneth Boulding suggested: "The idea of progress always precedes development."[44] This is merely a positive restatement of Bacon's observation that "the greatest obstacle to the advancement of the sciences . . . is to be found in men's despair and the idea of impossibility."[45] The *idea* of impossibility can be imposed by the limits of our worldview; the idea is not the same thing as impossibility itself. That is, the domain of the possible can be effectively narrowed by the myths and maps that guide our actions. Boulding argued: "Values are the food of knowledge and knowledge like any other organism moves toward that part of a possible field of growth where the values are the highest."[46] A different worldview—different paradigms, different myths—may create new maps, new topographies of value toward which knowledge may grow. What were once ideas of impossibility may gradually become ideas of progress.

The influence of values on the direction of scientific and technological progress has been one of the principal themes of this book, and it is now necessary to confront directly the implications of this theme. The values that drive the R&D agenda are also driving humanity into something of a corner. Foremost among these values is the reverence for economic growth, and in this sense the myth of the endless frontier is essentially utopian because it helps sustain the belief—necessary for the economic health of the modern, high-technology state—that the potential for such growth is endless.

Suppose, however, that continued economic growth and dissemination of wealth faced intractable limitations and obstacles over the next century or so. It is not necessary that this actually be the case, only that we believe it to be the case or strongly suspect that it might be. Suddenly, the argument that economic growth is the only reasonable route to global well-being loses its strength, as does the certainty that our own prosperity can be indefinitely expanded or even maintained. Suddenly, the metaphor of an endless frontier of science and technology has less currency, less intuitive appeal and practical value.

Now one might reasonably argue, as many experts have, that the environmental dilemmas facing humanity are strong evidence that growth of economies, of material consumption, of waste, of population, of conflict, of environmental degradation, cannot long continue unabated without severely retarding the prospects for global human development—not to mention causing great change to the earth's ecosystems and atmospheric stability. But other experts argue, perhaps with equal reason, that science and technology, in partnership with the economic marketplace, are precisely the tools that will allow us to overcome these hurdles—that the earth's resource capacity will therefore never be exhausted, that there are no natural limits to the creation of wealth in the global economy, and that if only the nations of the world would engage in the appropriate economic behavior, they would all eventually and inevitably grow to look more or less the same. Neither position is provable because both require a predictive capability that is—and will remain—beyond the capacity of scientific methodologies.

Apart from technical arguments, there are strong political and emotional reasons to prefer the infinite-growth scenario. In particular, it reinforces the wisdom of the status quo, and is thus both flattering to our collective egos and undemanding of social or political change, while running moral interference for global economic inequity (after all, someone has to be the first to achieve affluence). But as a ruling paradigm, this scenario is unquestionably riddled with discrepancies. For example, dilemmas of economic growth and environmental preservation are real. In affluent America, do we prefer ancient forests or jobs for our loggers? In Africa, do we value the lives of elephants, gorillas, and rhinoceroses over the livelihoods of the impoverished poachers who survive by killing these animals? Do we prefer Amazonian rain forests or economic growth in

Brazil? Do we prefer to control the greenhouse gas content of the atmosphere or to allow China to pursue its economic ambitions by exploiting its huge, cheap coal reserves? Society generally tries to answer such questions by devising more epicycles: retrain loggers, put animals in preserves, promote ecotourism in the rain forest, export energy technologies to China, do more research. But judging by the continued acceleration of virtually all types of environmental degradation and waste generation on a global scale, and by the demographic reality that 95 percent of all population growth in the coming decades will occur in countries that are already poor and environmentally stressed (although often well-armed), perhaps we need to start thinking in terms of ellipses.

If we accept the view that nature imposes intrinsic constraints on the direction, magnitude, and character of societal growth, then the topography of values will change and knowledge will move in new directions. Accepting this view of nature does not require that it be entirely correct in an objective sense, only that it be at least as reasonable as the prevailing frame of reference, and—like the Copernican universe—that it offer a fresh perspective for reducing the discrepancies arising out of the prevailing myths and paradigms.

For several decades at least, some scientists and some policy makers have recognized that environmental quality and global human development cannot reasonably be considered as separate issues. Environmental degradation is both a consequence of and a contributor to poverty and conflict, while economic development rapidly leads to intensification of resource exploitation and waste generation. The search for a political and intellectual framework within which to understand these types of relationships and by which to guide both research and policy is perhaps best embodied by the concept of sustainable development, which was first given prominence in the 1987 report *Our Common Future*, issued by the World Commission on Environment and Development under the aegis of the United Nations. As defined in the report, sustainable development encompassed the seemingly modest goal of meeting the developmental "needs of the present without compromising the ability of future generations to meet their own needs."[47] The concept of sustainability, however, has since been adopted for many applications—for example, sustainable agriculture, sustainable forestry, and sustainable energy use—and its meaning may vary depending on the user. This lexical and conceptual malleability has led some to dismiss sustainability as a buzzword or a cliché or a meaningless shibboleth, but one might alternatively argue that what we have here is the basis for a unifying mythology. As imprecise—or all encompassing—as the word has come to be, a general commonality of intent does exist among most of those who use it: sustainability is an alternative to a mentality of infinite growth. It is an alternative to endless frontiers. Sustainability may have lost its value as a term of art but gained value as an organizing principle that can unify action along diverse fronts.

Sustainability may therefore offer an alternative framework—a new mythology—for evaluating the contribution of science and technology to human development and welfare. Application of this framework requires that certain articles of faith be scrutinized and perhaps rejected, starting with the synonymy between unconstrained growth (of knowledge, innovation, wealth) and progress. The compatibility of science, technology, and market economics has arisen from a permissive view of progress in which magnitude and efficiency of growth are more important than direction. Because all growth is highly valued, this permissiveness has often been confused with value neutrality.

Sustainability, in contrast, dictates that progress is directional and relative; magnitude of growth is necessarily a less significant metric of progress than equity of distribution; rising efficiency and productivity contribute to progress only when they are accompanied by a reduction in the intensity of human impact on the natural environment and thus on the future generations who will depend on that environment. While economic markets can, in theory, incorporate some environmental costs into the pricing structure,[48] they cannot possibly address the problem of sustainability head-on: the needs and demands of future generations cannot be reflected in the marketplace of the present, and the impact of today's activities on tomorrow's environment (including the social and environmental costs of global climate change or extreme economic inequity) cannot be predicted—and therefore priced—with any precision. Because the marketplace cannot enforce criteria of sustainability, these criteria must be applied at an earlier point on the path of progress; for example, through the evolving organization, priorities, and goals of the

R&D system.[49] The policy recommendations discussed above suggest one possible direction in which this evolution could occur. The most obvious place to start is by redressing the preposterous mismatch between the R&D agenda of the North and the development priorities of the South.

The current turmoil in the R&D system portends that society as a whole, and the research community in particular, may be susceptible to alternative mythologies, yet these alternatives can hardly be imposed on the consciousness of the system by fiat. Vannevar Bush's "endless frontier" reflected the mood, the vision, the hopes, and the self-interest of scientists and policy makers at the end of World War II; the trick is to find a conceptual framework that is more appropriate to the needs and aspirations—and the reality—of the world today. "Sustainability" was appropriated, almost overnight, by many scientists, engineers, policy makers, development experts, members of the press, and even the public, and applied to a wide variety of economic, environmental, and societal problems, because it filled a conceptual vacuum. It encompassed a complex problem or set of problems that had previously seemed diffuse, and suggested a topography of value toward which new knowledge could migrate. These may be the qualities necessary to displace the endless frontier as the ruling myth and metaphor of science and technology policy, and inspire the pursuit of a greater convergence between progress in the laboratory and progress in the complex world of human beings.

NOTES

Chapter One: The End of the Age of Physics

1. *National Patterns of R&D Resources: 1992*, NSF 92–330 (Washington, D.C.: National Science Foundation, 1992), p. 46; Office of Management and Budget, *The Budget of the United States, Fiscal Year 1996* (Washington, D.C.: Executive Office of the President, 1995), part 1, pp. 94–95.
2. Leon M. Lederman, "The Advancement of Science," *Science* 256 (May 22, 1992): 1123.
3. Nathan Rosenberg, *Exploring the Black Box: Technology, Economics, and History* (New York: Cambridge University Press, 1994), chapter 11.
4. William L. Laurence, *Men and Atoms* (New York: Simon and Schuster, 1959).
5. This is not to say that natural scientists have accepted this debunkery without a fight; for example, see *Daedalus* 107 (Spring 1978), devoted to "Limits of Scientific Inquiry"; Richard Q. Elvee, ed., *Nobel Conference XXV: The End of Science?* (Saint Peter, Minn.: Gustavus Adolphus College, 1989); and Paul R. Gross and Norman Levitt, *Higher Superstition: The Academic Left and Its Quarrels with Science* (Baltimore, Md.: Johns Hopkins University Press, 1994).

Chapter Two: The Myth of Infinite Benefit

1. For a discussion of the political origins of the report, see Daniel J. Kevles, "The National Science Foundation and the Debate over Postwar Research Policy, 1942–1945," *Isis* 68 (1977): 5–26; and J. Merton England, "Dr. Bush Writes a Report: 'Science—The Endless Frontier,'" *Science* 191 (January 9, 1976): 41–47.
2. Vannevar Bush, *Science, the Endless Frontier* (Washington, D.C.: Office of Scientific Research and Development, 1945; reprint, Washington, D.C.: National Science Foundation, 1960), p. 5.
3. William J. Clinton and Albert Gore, Jr., *Science in the National Interest* (Washington, D.C.: Executive Office of the President, August 1994), pp. 1–2.
4. John R. Steelman, *Science and Public Policy* (Washington, D.C.: U.S. Government Printing Office, 1947), vol. 1, p. 6.
5. National Science Board, *Science and Engineering Indicators—1993*, NSB 93-1 (Washington, D.C.: U.S. Government Printing Office, 1993), pp. 331–34, 363; Office of Management and Budget, *The Budget of the United States Government, Fiscal Year 1992* (Washington, D.C.: Executive Office of the President, 1991), Historical Tables, pp. 37–48.

6. Office of Management and Budget, *The Budget of the United States Government, Fiscal Year 1993* (Washington, D.C.: Executive Office of the President), part 1, p. 90; Clinton and Gore, *Science in the National Interest*; Frank Press, "Science and Technology Policy for the Post–Vannevar Bush Era," in *Science and Technology Policy Yearbook*, ed. Albert H. Teich, Stephen D. Nelson, and Celia McEnaney (Washington, D.C.: American Association for the Advancement of Science, 1993), p. 9; *Report of the FASEB Consensus Conference on FY 1994 Federal Biomedical Research Funding* (Bethesda, Md.: Federation of American Societies for Experimental Biology, November 4–6, 1992), p. 4; Leon M. Lederman, "The Advancement of Science," *Science* 256 (May 22, 1992): 1122.

7. Testimony of Theodore Cooper, chairman and CEO of the Upjohn Company, in Committee on Science, Space, and Technology, *Renewing U.S. Science Policy: Private Sector Views* (Washington, D.C.: U.S. Government Printing Office, September 24, 1992), p. 23.

8. Steelman, *Science and Public Policy*, p. 4.

9. Leon Lederman, "Science, the End of the Frontier?" a report to the Board of Directors of the American Association for the Advancement of Sciences, January 1991, p. 4.

10. Ibid., p. 18.

11. Bush, *Science, the Endless Frontier*, p. 5. Rooted as the myth of inevitable benefit is in Vannevar Bush's report, it must be noted that time has proven several of Bush's fundamental assumptions to be wrong. For example, the postwar economic development of Japan repudiates his assertion that *"[a] nation which depends upon others for its new basic scientific knowledge will be slow in its industrial progress and weak in its competitive position in world trade, regardless of its mechanical skill."* (Emphasis in original; p. 19.)

12. For example, *1994 Economic Report of the President* (Washington, D.C.: U.S. Government Printing Office, 1994); U.S. Congress, Office of Technology Assessment, *Competing Economies: America, Europe and the Pacific Rim*, OTA-ITE-498 (Washington, D.C.: U.S. Government Printing Office, October 1991); Marc L. Miringoff, *The Index of Social Health, Monitoring the Social Well-Being of the Nation* (Tarrytown, N.Y.: Fordham Institute for Innovation in Social Policy, 1992); "Families on a Treadmill: Work and Income in the 1980s," a staff study prepared for the use of the members of the Joint Economic Committee, U.S. Congress, January 1992; Juliet Schor, *The Overworked American: The Unexpected Decline of Leisure* (New York: Basic Books, 1991).

13. National Science Board, *Science and Engineering Indicators—1993*, pp. 351–54.

14. For international comparisons of research expenditures, see National Science Foundation, *International Science and Technology Update: 1991*, NSF 91–309 (Washington, D.C.: 1991) and National Science Board, *Science and Engineering Indicators—1993*, pp. 375–79. For international health comparisons, see Public Health Service, *United States Health, 1990* (Washington, D.C.: U.S. Department of Health and Human Services, 1990); United Nations Development Programme, *Human Development Report 1994* (New York: Oxford University Press, 1994); World Bank, *World Development Report 1994* (New York: Oxford University Press, 1994).

15. Zvi Griliches, "R&D and Productivity: Measurement Issues and Economic Results," *Science* 237 (July 3, 1987): 31; also see Edwin Mansfield, "Academic Research and Industrial Innovation," *Research Policy* 20 (1991): 1–12; and "A Review of Edwin Mansfield's Estimate of the Rate of Return from Academic Research and Its Relevance to the Federal Budget Process," Congressional Budget Office Staff Memorandum, April 1993.

16. For example, Griliches, "R&D and Productivity"; H. A. Averch, "Notes on the Economics of R&D and Public Policy, 1985–1990," unpublished contractor report for the Office of Technology Assessment, 1991.

17. U.S. Congress, Office of Technology Assessment, *Pharmaceutical R&D: Costs, Risks and Rewards*, OTA-H-522 (Washington, D.C.: U.S. Government Printing Office, February 1993), p. 2.

18. Quoted in Wil Lepkowski, "Science-Technology Policy Seems Set for New Directions in Clinton Era," *Chemical and Engineering News* (December 7, 1992): 9.

19. J. Michael Bishop, "Paradoxical Stress: Science and Society in 1993," John P. McGovern Lecture on Science and Society, delivered at the February 1993 annual meeting of Sigma Xi, San Francisco, Calif., p. 13.

20. Daniel Kleppner, professor of physics at MIT, quoted in "Roundtable: Science under Stress," *Physics Today* (February 1992): 38–39.

21. "By any measure, the United States is the world leader in science" (Frank Press, "Talking our Way into Scientific Decline," speech presented at the National Academy of Sciences' 128th annual meeting, Washington, D.C., April 30, 1991, p. 2); "The research and higher education system in the United States is the envy of the world" (U. S. Congress, Office of Technology Assessment, *Federally Funded Research: Decisions for a Decade*, OTA-SET-490 [Washington, D.C.: U.S. Government Printing Office, May 1991], p. 3).

22. John M. Rowell, "Condensed Matter Physics in a Market Economy," *Physics Today* (May 1992): 44.

23. John A. Armstrong, "University Research: New Goals, New Practices," *Issues in Science and Technology* (Winter 1992–93): 50–53 (emphasis in original; Armstrong was vice president for science and technology at IBM).

24. Office of Technology Assessment, *Federally Funded Research,* p. 219.

25. For a detailed discussion of the many factors contributing to increased research expenditures and the growing population of researchers, see ibid., chapters 6 and 7; also see David Goodstein, "Scientific Elites and Scientific Illiterates," *Vital Speeches of the Day* (April 15, 1993): 403–8.

26. For example, Barbara J. Culliton, "Biomedical Funding: The Eternal 'Crisis,'" *Science* 250 (December 21, 1990): 1652–53.

27. For example, "The Future Supply of Natural Scientists and Engineers," in *The State of Academic Science and Engineering,* NSF 90–35 (Washington, D.C.: National Science Foundation, 1990), pp. 189–232.

28. "The Ph.D. Shortage: The Federal Role," a policy statement of the Association of American Universities, January 11, 1990, pp. iii–iv (emphasis in original).

29. Richard C. Atkinson, "Supply and Demand for Scientists and Engineers: A National Crisis in the Making," *Science* 246 (April 27, 1990): 425–32.

30. Testimony of Dr. Philip Schambra, director, Fogarty International Center, National Institutes of Health, in Committee on Science, Space, and Technology, *Increasing U.S. Scientific Manpower* (Washington, D.C.: U.S. Government Printing Office, July 31, 1990), p. 20.

31. Testimony of Dr. Phillip A. Griffiths, provost, Duke University, in ibid., p. 38.

32. Dana Milbank, "Shortage of Scientists Approaches a Crisis as More Students Drop out of the Field," *Wall Street Journal,* September 17, 1990, E1.

33. For example, Christine M. Matthews, "Scientific Personnel: Supply and Demand," *Report for Congress* 92–419 SPR (Washington, D.C.: Congressional Research Service, May 8, 1992).

34. For example, John Markoff, "A Corporate Lag in Research Funds Is Causing Worry," *New York Times,* January 22, 1990, 1.

35. From an undated Young Scientists' Network (YSN) Newsletter. The YSN is a "no membership fee group which is mostly linked by computer electronic mail."

36. Malcolm Browne, "Amid 'Shortage,' Young Physicists See Few Jobs," *New York Times,* March 10, 1992, C1. A year later, the *Times* ran a veritable obituary for the scientific job market: Peter Kilborn, "The Ph.D.'s Are Here, But the Lab Isn't Hiring," *New York Times,* July 18, 1993, E3.

37. All the same, some young scientists, having spent years in training only to find themselves unable to land a good position, see darker motives behind this phenomenon. One physicist points to "senior members of the profession

who are either tenured professors or have good, secure positions. Encouraging more students to go into physics increases the supply of cheap labor that would enable their own groups to obtain more research funding, and increases membership in their community, with all its trimmings (for them)." V. K. Yee, letter to *Physics Today* (July 1993): 11.

38. Committee on Science, Space, and Technology, *Projecting Science and Engineering Requirements for the 1990s: How Certain Are the Numbers?* (Washington, D.C.: U.S. Government Printing Office, April 8, 1992).

39. For example: Robert M. White, "Too Many Researchers, Too Few Dollars," *Issues in Science and Technology* (Spring 1991): 35–37; Alan Fechter, "Engineering Shortages and Shortfalls: Myths and Realities," *The Bridge* 20 (Fall 1990): 16–20; U.S. Congress, Office of Technology Assessment, *Educating Scientists and Engineers: Grade School to Grad School*, OTA-SET-377 (Washington, D.C.: U.S. Government Printing Office, June 1988).

40. White, "Too Many Researchers," p. 35.

41. Zachary Burton, "U. S. Gets Its Money's Worth in Science," letter to *New York Times*, July 12, 1994, A12. The author is associate professor of biochemistry at Michigan State University.

42. For example, Lederman, "Science, the End of the Frontier?"

43. Ralph Gomery, president of the Alfred P. Sloan Foundation and formerly the senior vice president for science and technology at IBM, quoted in "Roundtable: Physics in Transition," *Physics Today* (February 1993): 40.

44. Steelman, *Science and Public Policy*, p. viii.

45. For example, "Culture Wars," letters to *Science* 265 (August 12, 1994): 853–54.

Chapter Three: The Myth of Unfettered Research

1. Ernest L. Eliel, *Science and Serendipity, The Importance of Basic Research* (Washington, D.C.: American Chemical Society, March 1993), p. 2.

2. Most of this increase occurred during the 1960s and 1980s; since 1953, government support for basic research grew faster than inflation every year except for the period 1968–75. *National Patterns of R&D Resources: 1992*, NSF 92–330 (Washington, D.C.: National Science Foundation, 1992), p. 50; Office of Management and Budget, *The Budget of the United States, Fiscal Year 1996* (Washington, D.C.: Executive Office of the President, 1995), part 1, pp. 94–95.

3. Vannevar Bush, *Science, the Endless Frontier* (Washington, D.C.: Office of Scientific Research and Development, 1945; reprint, Washington, D.C.: National Science Foundation, 1960), pp. 18–19.

4. National Science Board, *Science and Engineering Indicators 1993*, NSB 93–1 (Washington, D.C.: U.S. Government Printing Office, 1993), p. 347.

5. Arthur Kornberg, "Understanding Life as Chemistry," *Clinical Chemistry* 37 (1991): 1896.

6. William F. Raub, "Investigator-Initiated Research—The Productive Paradox," *Bioscience* 42 (July/August 1992): 550–55.

7. Daniel Kleppner, "The Mismeasure of Science," *The Sciences* (May/June 1991): 18–21.

8. Bush, *Science, the Endless Frontier,* p. 12.

9. The social context for scientific research, and the influence of this context on research outcomes, has been a subject of scholarly research in the social sciences for the last several decades. For some recent contributions, see Sheila Jasanoff, Gerald E. Markle, James C. Petersen, and Trevor Pinch, eds., *Handbook of Science and Technology Studies* (London: Sage Publications, 1995); Susan E. Cozzens and Thomas R. Gieryn, eds., *Theories of Science in Society* (Bloomington: Indiana University Press, 1990).

10. Testimony of Richard A. Muller, in Committee on Science, Space, and Technology, *Adequacy, Direction, and Priorities for the American Science and Technology Effort* (Washington, D.C.: U.S. Government Printing Office, February 28–March 1, 1989), p. 132. Perhaps the most influential formal statement of this doctrine can be found in: Michael Polanyi, "The Republic of Science: Its Political and Economic Theory," *Minerva* 1 (Autumn 1962): 54–73.

11. "[Basic] biological research benefits the nation's health, social well-being, and its economy. Basic research enables us to increase our understanding about life processes and contributes to the development of more effective medical treatments including new drugs and biotechnology products." (*Report of the FASEB Consensus Conference on FY 1994 Federal Biomedical Research Funding* [Bethesda, Md.: Federation of American Societies for Experimental Biology, November 4–6, 1992], p. 4.)

12. National Science Board, *Science and Engineering Indicators—1993*, pp. 351–54.

13. There is a substantial literature on the relationship between scientific research and technological innovation. For a particularly entertaining and accessible account, see Deborah Shapley and Rustum Roy, *Lost at the Frontier: U.S. Science and Technology Policy Adrift* (Philadelphia: iSi Press, 1985); for a more technical summary, see Stephen J. Kline and Nathan Rosenberg, "An Overview of Innovation," in *The Positive Sum Strategy, Harnessing Technology for Economic Growth,* ed. Ralph Landau and Nathan Rosenberg (Washington, D.C.: National Academy Press, 1986), pp. 275–305.

14. Eliel, *Science and Serendipity*; three articles by Patrick J. Hannan, Rustum Roy, and John F. Christman, "Prince Serendip at Work," *Chemtech* (January 1988): 18–21; "Chance and Drug Discovery," *Chemtech* (February 1988): 80–

83; "Serendipity in Chemistry, Astronomy, Defense, and Other Useless Fields," *Chemtech* (July 1988): 402–6; and "Wasted Dollars," *The Economist* (September 3, 1988): 64.

15. Donald E. Stokes, *Pasteur's Quadrant—Basic Science and Technological Innovation* (Washington, D.C. Brookings Institution, forthcoming).(Reference is to manuscript in preparation, January 10, 1995, draft, chapter 1, p. 15.)

16. John M. Rowell, "Condensed Matter Physics in a Market Economy," *Physics Today* (May 1992): 46.

17. For example, see Shapley and Roy, *Lost at the Frontier;* and National Academy of Sciences, *Applied Science and Technological Progress: A Report to the U.S. House of Representatives Committee on Science and Astronautics* (Washington, D.C.: U.S. Government Printing Office, 1967).

18. Testimony of Dr. Arthur L. Schawlow, in Committee on Science and Technology, *Views on Science Policy of the American Nobel Laureates for 1981* (Washington, D.C.: U.S. Government Printing Office, February 25, 1982), p. 6 (emphasis added).

19. William O. Baker, "Response," in Shapley and Roy, *Lost at the Frontier,* pp. 161–67; Howard E. Simmons, "Basic Research—A Perspective," *Chemical and Engineering News* (March 14, 1994): 27–31.

20. Michael Schrage, "Physicists' Reign Is Likely to End," *Los Angeles Times,* October 3, 1991, D1.

21. George E. Brown and Radford Byerly, Jr., "Research in EPA: A Congressional Point of View," *Science* 211 (March 27, 1981): 1385–90; Carnegie Commission on Science, Technology, and Government, *Environmental Research and Development: Strengthening the Federal Infrastructure* (New York: Carnegie Commission on Science, Technology, and Government, December 1992).

22. National Science Board, *Science and Engineering Indicators—1993*, pp. 351–52.

23. Daniel J. Kevles, *The Physicists: The History of a Scientific Community in Modern America* (Cambridge, Mass.: Harvard University Press, 1987), pp. 345–46. Kevles quotes from a 1946 letter written by Vannevar Bush; also see Daniel J. Kevles,"The National Science Foundation and the Debate over Postwar Research Policy, 1942–1945," *Isis* 68 (1977): 5–26.

24. Nobel prize–winning economist Herbert Simon writes: "In a word, the physical and biological worlds are not the problem; we are the problem. It is time that we took seriously the ancient injunction to know ourselves. Knowing ourselves will not solve our human problems magically, but it will open new paths, unknown to us now, for progressing toward a better world." (*The Objectivity Crisis: Rethinking the Role of Science in Society,* Chairman's Report to the Committee on Science, Space, and Technology [Washington, D.C.: U.S. Government Printing Office, 1993], p. 17–18.)

25. "[The] men directing national policy—especially those deriving from Harvard University and MIT—believed that the ultimate test of American society in its competition with the Soviets boiled down to finding out which contestant could develop superior skills in every field of human endeavor. . . . But success would come only . . . if it were encouraged by the removal of long-standing fiscal limitations on education, research, and development." (William H. McNeill, *The Pursuit of Power* [Chicago: University of Chicago Press, 1982], pp. 368–69.)

26. National Science Board, *Science and Engineering Indicators—1993*, pp. 78–80; U.S. Congress, Office of Technology Assessment, *Educating Scientists and Engineers: Grade School to Grad School*, OTA-SET-377 (Washington, D.C.: U.S. Government Printing Office, June 1988), chapter 3; Marguerite Holloway, "A Lab of Her Own," *Scientific American* (November 1993): 72–79.

27. Harriet Zuckerman, "The Careers of Men and Women Scientists: A Review of Current Research," in *The Outer Circle: Women in the Scientific Community*, ed. Harriet Zuckerman, Jonathan R. Cole, and John T. Bruer (New York: W. W. Norton, 1991), p. 47.

28. Quoted in Marcia Barinaga, "Feminists Find Gender Everywhere in Science," *Science* 260 (April 16, 1993): 393. The source of the quotation was a "physicist familiar with the feminist literature who requested anonymity."

29. Carol Taylor, "Gender Equity in Research," *Journal of Women's Health* 3 (1994): 143–53; Susan Wood, "Issues in Women's Health Research," in *Science and Technology Policy Yearbook, 1992*, ed. Stephen D. Nelson, Kathleen M. Gramp, and Albert H. Teich (Washington, D.C.: American Association for the Advancement of Science, 1993), pp. 315–18; Paul Cotton, "Is There Still Too Much Extrapolation from Data on Middle-Aged White Men?" *Journal of the American Medical Association* 263 (February 23, 1990): 1049–50; Vivian W. Pinn, "Commentary: Women, Research, and the National Institutes of Health," *American Journal of Preventative Medicine* 8 (1992): 324–27; and papers from the session "Sex Bias in Research: Are Males and Females the Same?" Annual Meeting of the American Association for the Advancement of Science, Boston, Mass., February 1993.

30. For example, Taylor, "Gender Equity"; Pinn, "Commentary."

31. Taylor, "Gender Equity"; Vivian W. Pinn, "Women's Health Research," *Journal of the American Medical Association* 268 (October 14, 1992): 1921–22. Most influential was a report issued by the General Accounting Office of the U.S. Congress: *National Institutes of Health: Problems in Implementing Policy on Women in Study Populations*, GAO/T-HRD-90-38 (Washington, D.C.: U.S. Government Printing Office, 1990).

32. Bernadine Healy, quoted in "U.S. Health Study to Involve 160,000 Women at 16 Centers," *New York Times*, March 31, 1993, A21.

33. Marcia Barinaga, "Is There a 'Female Style' in Science?" *Science* 260 (April 16, 1993): 384–91; Holloway, "A Lab of Her Own"; Jeffrey Mervis, "Radcliffe President Lambastes Competitiveness in Research," *The Scientist* (January 20, 1992): 3, 7; Linda Wilson, "U.S. Research Universities Now Confront Fateful Choices," *The Scientist* (March 16, 1992): 11. Clifford Adelman, a researcher with the U.S. Department of Education, analyzed data on women's academic and job performance and concluded that "women's aspirations are less inflated than men's, their plans more realistic, their focus on goals more intense. . . . [Women are] also more willing to share their knowledge to the benefit of the organization." (From "Women's Century," unpublished ms., 1990, p. 4.)

34. Evelyn Fox Keller, *A Feeling for the Organism* (New York: W. H. Freeman, 1983); Nobel Assembly of the Karolinsk Institute, "Press Release, October 10, 1983," reprinted in Committee on Science and Technology, *Views on Science Policy of the American Nobel Laureates for 1983* (Washington, D.C.: U.S. Government Printing Office, March 8, 1984), pp. 32–39; Ruth Hubbard, *The Politics of Women's Biology* (New Brunswick, N.J.: Rutgers University Press, 1990), pp. 52–65.

35. Keller, *A Feeling for the Organism*.

36. Ibid., chapter 12.

37. Another example might be the thirty-year dominance of the field of primatology by three now-famous female scientists, Jane Goodall, the late Dian Fossey, and Biruté Galdikas.

38. Evelyn Fox Keller, "The Wo/Man Scientist: Issues of Sex and Gender in the Pursuit of Science," in Zuckerman, Cole, and Bruer, *The Outer Circle*, p. 234.

39. Shirley M. Tilghman, "Science vs. Women—A Radical Solution," *New York Times*, January 26, 1993, A17.

Chapter Four: The Myth of Accountability

1. Gerald Holton, "The Value of Science at the 'End of the Modern Era,'" speech delivered at February 1993 annual meeting of Sigma Xi, San Francisco, Calif., pp. 4–5.

2. Ibid., p. 15.

3. Ibid., p. 5.

4. Paul Kengor and Peter D. Zimmerman, "Science and Citizenship," *Christian Science Monitor*, June 21, 1993, 19.

5. Paul R. Ehrlich, "One Ecologist's Opinion on the So-Called Stanford Scandals and Social Responsibility," *BioScience* 42 (October 1992): 702.

6. David Baltimore, quoted in Pamela S. Zurer, "Scientific Whistleblower Vindicated," *Chemical and Engineering News* (April 8, 1991): 36.

7. Sherwood F. Rowland, "President's Lecture: The Need for Scientific Communication with the Public," *Science* 260 (June 11, 1993): 1573. Rowland implies that such distinctions can be made by those with adequate "understanding." This misconception is common among scientists; see Chapter 5.

8. Sheldon Glashow, "The Death of Science!?" in *Nobel Conference XXV: The End of Science?* ed. Richard Q. Elvee (Saint Peter, Minn.: Gustavus Adolphus College, 1989), p. 24.

9. National Science Board, *Science and Engineering Indicators—1991*, NSB 91-1 (Washington, D.C.: U.S. Government Printing Office, 1991), pp. 455, 464–65.

10. Walter E. Massey, "Science Education in the United States: What the Scientific Community Can Do," *Science* 245 (September 1, 1989): 915–21.

11. Science literacy was defined by surveyors as the ability to "combine a general understanding of how science is conducted with a minimal knowledge of scientific concepts and an understanding of how science impacts society and the daily lives of individuals." ("No Gains in U.S. Scientific Literacy, New Survey Shows," *National Science Foundation News*, NSF PR 89–4 [January 18, 1989].)

12. National Science Board, *Science and Engineering Indicators—1991*, p. 467.

13. Ibid., p. 464 (emphasis added).

14. National Science Board, *Science and Engineering Indicators—1993*, NSB 93-1 (Washington, D.C.: U.S. Government Printing Office, 1993), pp. 485, 489.

15. Bernhard Zechendorf, "What the Public Thinks about Biotechnology," *Bio/Technology* 12 (September 1994): 870–75.

16. National Science Board, *Science and Engineering Indicators—1993*, p. 483.

17. For example, compare Ehrlich, "One Ecologist's Opinion," and Frederick Grinnell, *The Scientific Attitude* (New York: Guilford Press, 1992), chapter 6.

18. Panel on Scientific Responsibility and the Conduct of Research, *Responsible Science: Ensuring the Integrity of the Research Process* (Washington, D.C.: National Academy Press, 1992); also see Bruce Alberts and Kenneth Shine, "Scientists and the Integrity of Research," *Science* 266 (December 9, 1994): 1660–61.

19. Sheila Widnall, "Fostering Scientific Integrity," in *Science and Technology Policy Yearbook, 1992*, ed. Stephen D. Nelson, Kathleen M. Gramp, and Albert H. Teich (Washington, D.C.: American Association for the Advancement of Science, 1993), p. 16.

20. Holton, "The Value of Science," p. 8.

21. This is a somewhat unfortunate analogy, forced upon the author by the

previous citation. There is, of course, no intent to compare the scientific community to the tobacco industry or the National Rifle Association, except insofar as each represents a relatively insular subculture.

22. Of fifteen nations surveyed, five of the six nations with the highest literacy rates also had the lowest rates of agreement with the following statements: "Scientists can be trusted to make the right decisions," and "The benefits of science are greater than any harmful effects." National Science Board, *Science and Engineering Indicators—1991*, pp. 464, 469.

23. For an engaging discussion of the gap between promise and performance in a particular discipline—molecular genetics—see Ruth Hubbard and Elijah Wald, *Exploding the Gene Myth: How Genetic Information is Produced and Manipulated by Scientists, Physicians, Employers, Insurance Companies, Educators, and Law Enforcers* (Boston: Beacon Press, 1993).

24. "Hopeful Talk on Science as Press Leaves Academy," *Physics Today* (July 1993): 64.

25. Ibid.

26. Testimony of Robert A. Frosch, vice president of General Motors Research Laboratories, in Committee on Science, Space, and Technology, *Renewing U.S. Science Policy: Private Sector Views* (Washington, D.C.: U.S. Government Printing Office, September 24, 1992), p. 75. Also see Council on Competitiveness, *Gaining New Ground: Technology Priorities for America's Future* (Washington, D.C.: Council on Competitiveness, 1991).

27. World Bank, *World Development Report 1993* (New York: Oxford University Press, 1993).

28. Jon Cohen, "Somber News from the AIDS Front," *Science* 260 (June 18, 1993): 1712–13; Malcolm Gladwell, "Meeting Shows Wide Gap in Knowledge about AIDS," *Washington Post*, July 26, 1992, A26.

29. Leslie Roberts, "Learning from an Acid Rain Program," *Science* 251 (March 15, 1991): 1302–5.

30. Edward Rubin, Lester B. Lave, and M. Granger Morgan, "Keeping Climate Research Relevant," *Issues in Science and Technology* (Winter 1991–92): 48–49.

31. Ibid., p. 49.

32. Committee on Recombinant DNA Molecules, "Potential Biohazards of Recombinant DNA Molecules," letter to *Science* 185 (July 26, 1974): 303; also see Sheldon Krimsky, *Genetic Alchemy: The Social History of the Recombinant DNA Controversy* (Cambridge, Mass.: MIT Press, 1982).

33. David Baltimore, "Limiting Science: A Biologist's Perspective," *Daedalus* 107 (Spring 1978): 39.

34. From an essay by David Pramer and Janet Shoemaker of the American So-

ciety for Microbiology, cited in Bernard Dixon, "Don't Believe (All) the Hype," *Bio/Technology* 12 (June 1994): 555.

35. Maxine Singer and Dieter Soll, "Guidelines for DNA Hybrid Molecules," letter to *Science* 181 (September 21, 1973): 1114.

36. Susan Hassler, "Not Science, But Necessary," *Bio/Technology* 12 (January 12, 1994): 7.

37. Irene Stith-Coleman, "Human Fetal Research and Tissue Transplantation," *Issue Brief* IB88100 (Washington, D.C.: Congressional Research Service, December 28, 1989).

38. Joseph Palca, "Fetal Tissue Transplants Remain Off Limits," *Science* 246 (November 10, 1989): 752.

39. Philip J. Hilts, "Citing Abortion, U.S. Extends Ban on Grants for Fetal Tissue Work," *New York Times*, November 2, 1989, 1, B19.

40. Ibid.

41. Palca, "Fetal Tissue."

42. Howard W. Jones, Jr., "Needless Infertility," *Headline News, Science Views*, ed. David Jarmul (Washington, D.C.: National Academy Press, 1991), pp. 89–91.

43. David Beckler of the Carnegie Commission on Science, Technology, and Government, quoted in Wil Lepkowski, "Science-Technology Policy Seems Set for New Directions in Clinton Era," *Chemical and Engineering News* (December 7, 1992): 7–14.

44. For example, see Joseph Palca, "Scientists Take One Last Swing," *Science* 257 (July 3, 1992): 20–21; also see Lepkowski, "Science-Technology Policy Seems Set for New Directions."

45. For example, see Arthur Kornberg, "Science Is Great, But Scientists Are Still People," *Science* 257 (August 14, 1992): 859.

46. Lucy Shapiro, Stanford University molecular biologist, quoted in Jeffrey Mervis, "U.S. Research Forum Fails to Find a Common Front," *Science* 263 (February 11, 1994): 752.

47. Daniel Kleppner, "Thoughts on Being Bad," *Physics Today* (August 1993): 9.

Chapter Five: The Myth of Authoritativeness

1. Daniel E. Koshland, Jr., "Two Plus Two Equals Five," *Science* 247 (March 23, 1990): 1381.

2. Harold D. Lasswell, *Psychopathology and Politics* (Chicago: University of Chicago Press, 1977): 184.

3. Susan Watts, "Science Advice: An Abuser's Guide," *New Scientist* (March 10, 1990): 55–59.

4. Some examples include: Irwin Goodwin, "Happer Leaves DoE under Ozone

Cloud for Violating Political Correctness," *Physics Today* (June 1993): 89–91; Philip Shabecoff, "Scientist Says Budget Office Altered His Testimony," *New York Times,* May 8, 1993, A1; William K. Stevens, "What Really Threatens the Environment," *New York Times,* January 29, 1991, C4; Malcolm Gladwell, "Pediatric AIDS Studied at Adults' Expense," *Washington Post,* October 5, 1992, A1, A10.

5. Letter to the Chair of the House of Representatives Committee on Science, Space, and Technology (name withheld to preserve confidentiality).

6. Carnegie Commission on Science, Technology, and Government, *Science, Technology, and Congress: Expert Advice and the Decision-Making Process* (New York: Carnegie Commission on Science, Technology, and Government, February, 1991), pp. 8, 9, 11. For other approaches, see John Doble and Amy Richardson, "You Don't Have to be a Rocket Scientist," *Technology Review* (January 1992): 51–54; John F. Ahearne, "Answering Public Concerns in Science," *Physics Today* (September 1988): 36–42.

7. Koshland, "Two Plus Two."

8. Thomas H. Kean, "Making the Link between Science and Politics," in *Headline News, Science Views,* ed. David Jarmul (Washington, D.C.: National Academy Press, 1991), pp. 16–18. Kean was governor of New Jersey when he wrote this op-ed piece.

9. George C. Giddings, "Politics and Science," letter to *Chemical and Engineering News* (July 19, 1993): 4–5. Giddings described himself as a "30-year veteran of all aspects of food irradiation."

10. Leon Lederman, "Physics for Poets, Science for Society," in Jarmul, *Headline News,* pp. 11–13.

11. Just about any scientist who has worked as an advisor to a politician knows this to be true. For example, Edward E. David, "White House Science Advising," in *Science and Technology Advice to the President, Congress, and Judiciary,* ed. William T. Golden (Washington, D.C.: AAAS Press, 1993), pp. 104–11.

12. Lasswell, *Psychopathology and Politics,* pp. 184–85.

13. For example, see Harvey Brooks, "Expertise and Politics—Problems and Tensions," *Proceedings of the American Philosophical Society* 119 (August 1975): 257–61.

14. "When an issue becomes highly controversial—when it is surrounded by uncertainties and conflicting values—then expertness is very hard to come by, and it is no longer easy to legitimate the experts. In these circumstances, we find that there are experts for the affirmative and experts for the negative. We cannot settle such issues by turning them over to particular groups of experts. At best, we may convert the controversy into an adversary proceeding in

which we, the laymen, listen to the experts but have to judge between them." Herbert A. Simon, *Reason in Human Affairs* (Stanford, Calif.: Stanford University Press, 1983), p. 97.

15. A series of editorials and response letters in *Science* outlines some of the terms of debate. See, for example: Philip H. Abelson, "Pesticides and Food," *Science* 259 (February 26, 1993): 1235, and response letters in *Science* 260 (June 4, 1993): 1409–10; also Philip H. Abelson, "Pathological Growth of Regulations," *Science* 260 (June 25, 1993): 1859, and response letters in *Science* 261 (September 10, 1993): 1373–75.

16. Harold D. Lasswell and Abraham Kaplan, *Power and Society: A Framework for Political Inquiry* (New Haven, Conn.: Yale University Press, 1950), pp. xv–xvi.

17. For example, see Ronald D. Brunner and William Ascher, "Science and Social Responsibility," *Policy Sciences* 25 (1992): 295–331; and William Ascher, "The Forecasting Potential of Complex Models," *Policy Sciences* 13 (1981): 247–67.

18. Stephanie Pain, "Valdez Spill Wasn't So Bad, Claims Exxon," *New Scientist* (May 8, 1993): 4.

19. Richard Stone, "Dispute over Exxon Valdez Cleanup Gets Messy," *Science* 260 (May 7, 1993): 749.

20. *Daubert v Merrell Dow Pharmaceuticals, Inc.,* 113 S. Ct. 2786 (1993); Jeffrey Mervis, "Supreme Court to Judges: Start Thinking Like Scientists," *Science* 261 (July 2, 1993): 22; Eliot Marshall, "Supreme Court to Weigh Science," *Science* 259 (January 29, 1993): 588–90.

21. R. C. Lewontin and Daniel L. Hartl, "Population Genetics in Forensic DNA Typing," *Science* 254 (December 20, 1991): 1745–50.

22. Leslie Roberts, "Fight Erupts over DNA Fingerprinting," *Science* 254 (December 20, 1991): 1721–23; "*Science* Editor Denies Yielding to FBI Pressure," *Science and Government Report* 22 (February 15, 1992): 1–3.

23. Jane Kirtley, executive director of the Reporters Committee for Freedom of the Press, quoted in "*Science* Editor Denies Yielding," p. 2.

24. Christopher Wills, "Forensic DNA Typing," letter to *Science* 255 (February 28, 1992): 1050 (emphasis in original).

25. Steven N. Austad, "Forensic DNA Typing," letter to *Science* 255 (February 28, 1992): 1050.

26. For a popular account of the dispute, see Edward Humes, "The DNA Wars," *Los Angeles Times Magazine,* November 29, 1992, 20–26, 54–57. Also see Rachel Nowak, "Forensic DNA Goes to Court with O.J.," *Science* 265 (September 2, 1994): 1352–54.

27. B. Devlin, Neil Risch, and Kathryn Roeder, "Statistical Evaluation of DNA Fingerprinting: A Critique of the NRC's Report," *Science* 259 (February 5, 1993): 748–49, 837.

28. Daniel L. Hartl and Richard C. Lewontin, "DNA Fingerprinting Report," letter to *Science* 260 (April 23, 1993): 473–74.

29. B. Devlin, Neil Risch, and Kathryn Roeder, "NRC Report on DNA Typing," letter to *Science* 260 (May 21, 1993): 1057–58.

30. Lynwood R. Yarbrough, "Forensic DNA Typing," letter to *Science* 255 (February 28, 1992): 1052.

31. Roberts, "Fight Erupts over DNA," p. 1721.

32. A geneticist "who wants to stay out of the fray and thus seeks anonymity," quoted in ibid.

33. Daniel E. Koshland, Jr., "The DNA Fingerprint Story (Continued)," *Science* 265 (August 19, 1994): 1015; Daniel L. Hartl and Richard C. Lewontin, "DNA Fingerprinting," letter to *Science* 266 (October 14, 1994): 201.

34. Watts, "Science Advice"; Lederman, "Physics for Poets."

35. For an early report on global change, see Assembly of Mathematical and Physical Sciences, *Energy and Climate* (Washington, D.C.: National Academy of Sciences, 1977).

36. Philip Shabecoff, "Global Warming Has Begun, Expert Tells Senate," *New York Times*, June 24, 1988, A1, A14.

37. Committee on Earth and Environmental Sciences, *Our Changing Planet: The FY 1992 U.S. Global Change Research Program, A Supplement to the U.S. President's Fiscal Year 1992 Budget* (Washington, D.C.: Committee on Earth and Environmental Sciences), p. 1. The seven "science priorities" are: Climate and Hydrologic Systems, Biogeochemical Dynamics, Ecological Systems and Dynamics, Earth System History, Human Interactions, Solid Earth Processes, and Solar Influences. About 70 percent of the USGCRP budget is devoted to the first two priorities.

38. Congressman Don Ritter, in Committee on Science, Space, and Technology, *U.S. Global Change Research Program* (Washington, D.C.: U.S. Government Printing Office, May 6, 1992), p. 5.

39. Committee on Earth and Environmental Sciences, *Our Changing Planet*, p. 52.

40. Ibid., p. 15.

41. For example, see: Naomi Oreskes, Kristin Shrader-Frechette, and Kenneth Belitz, "Verification, Validation, and Confirmation of Numerical Models in the Earth Sciences," *Science* 263 (February 4, 1994): 641–46; Richard A. Kerr, "Climate Modeling's Fudge Factor Comes under Fire," *Science* 265 (September 9, 1994): 1528; Mitchell M. Waldrop, *Complexity: The Emerging Science at the Edge of Order and Chaos* (New York: Simon and Schuster, 1992).

42. J. D. Mahlman, "Assessing Global Climate Change: When Will We Have Better Evidence?" in *Climate Change and Energy Policy*, LA-LR-92–502, ed.

Louis Rosen and Robert Glasser (Los Alamos, N.M.: Los Alamos National Laboratory, May 1992), pp. 317–31.

43. Robert Eisner, "Why the Long Faces?" *New York Times Book Review,* September 19, 1993, 18.

44. "Energy Survey, In the Engine Room," *The Economist* (June 18, 1994): 17.

45. Some effort has been devoted to visualizing a global-change research agenda that is more compatible with the needs of policy makers and the realities of both scientific progress and political action. For example, see: Ronald D. Brunner, "Global Climate Change: Defining the Policy Problem," *Policy Sciences* 24 (1991): 291–311; Radford Byerly, Jr., "The Policy Dynamics of Global Change," *EarthQuest* (Spring 1989): 11–13, 24; *Joint Climate Project to Address Decision Maker's Uncertainties, Report of Findings* (Washington, D.C.: Joint Climate Project, May 1992); and testimony in Committee on Science, Space, and Technology, *Global Change Research: Science and Policy* (Washington, D.C.: U.S. Government Printing Office, May 19, 1993).

46. Watts, "Science Advice"; Susan A. Edelman, "The American Supersonic Transport," in *The Technology Pork Barrel,* ed. Linda R. Cohen and Roger G. Noll (Washington, D.C.: Brookings Institution, 1991), pp. 97–137; Daniel J. Kevles, *The Physicists: The History of a Scientific Community in Modern America* (Cambridge, Mass.: Harvard University Press, 1987), pp. 410–12.

47. David E. Gushee, "Stratospheric Ozone Depletion," *Issue Brief* IB89021 (Washington, D.C.: Congressional Research Service, February 14, 1992); Richard Elliot Benedick, "Protecting the Ozone Layer: New Directions in Diplomacy," in *Preserving the Global Environment,* ed. Jessica Tuchman Mathews (New York: W. W. Norton, 1991), pp. 112–53.

48. Ibid.

49. Ibid.

50. Benedick, "Protecting the Ozone Layer," p. 129 (emphasis in original); also see: U.S. Congress, Office of Technology Assessment, *Changing by Degrees: Steps to Reduce Greenhouse Gases,* OTA-O-482 (Washington, D.C.: U.S. Government Printing Office, February 1991).

51. Benedick, "Protecting the Ozone Layer," p. 132.

52. Office of Technology Assessment, *Changing by Degrees;* Gushee, "Stratospheric Ozone Depletion."

53. Goodwin, "Happer Leaves DoE under Ozone Cloud."

54. Gary Taubes, "The Ozone Backlash," *Science* 260 (June 11, 1993): 1580–83.

55. Rudy Abramson, "Effort to Change Wetlands Definition Sparks Protests," *Los Angeles Times,* March 29, 1991, A4; Charles C. Mann, "Extinction: Are Ecologists Crying Wolf?" *Science* 253 (August 16, 1991): 736–38; U.S. Congress,

Office of Technology Assessment, *Energy Technology Choices: Shaping Our Future*, OTA-E-493 (Washington, D.C.: U.S. Government Printing Office, July 1991); Gary Lee, "Dioxin Study Spurs Plea for Restrictions," *Washington Post*, September 14, 1994, A8; Richard Stone, "Environmental Estrogens Stir Debate," *Science* 265 (July 15, 1994): 308–10; "Ground-Water Cleanup vs. Ground-Water Protection—Where Should the $$$ Go?" *GSA Today* 3 (July 1993): 179–84; "Major EMF Report Warns of Health Risks," *Science* 269 (August 18, 1995): 911.

56. Richard A. Kerr, "Scientists See Greenhouse, Semiofficially," *Science* 269 (September 27, 1995): 1667.

57. Ernest W. Volkmann, "Alar Controversy," letter to *Chemical and Engineering News* (December 4, 1989): 2–3.

Chapter Six: The Myth of the Endless Frontier

1. "Everyone knows that the linear model of innovation is dead. That model represented the innovation process as one in which technological change was closely dependent upon, and generated by, prior scientific research. It was a model that, however flattering it may have been to the scientist and the academic, was economically naive and simplistic to the extreme." Nathan Rosenberg, "Critical Issues in Science Policy Research," *Science and Public Policy* 18 (December 1991): 335. Also see, for example, Stephen J. Kline, "Innovation Is Not a Linear Process," *Research Management* 28 (July–August 1985): 36–45; George Wise, "Science and Technology," *Osiris*, 2d ser., 1 (1985): 229–46.

2. Robert K. Adair and Ernest M. Henley, "*Physical Review* Centenary—From Basic Research to High Technology," *Physics Today* (October 1993): 25.

3. For example, see Alvin M. Weinberg, "The Axiology of Science," *American Scientist* 58 (November–December 1970): 612–17.

4. Sheldon Glashow, "The Death of Science!?" in *Nobel Conference XXV: The End of Science?* ed. Richard Q. Elvee (Saint Peter, Minn.: Gustavus Adolphus College, 1989), p. 26.

5. Howard K. Birnbaum, "Searching for Science Criticism's Sources," letter to *Physics Today* (June 1994): 15.

6. Lynn White, Jr., *Machina Ex Deo: Essays in the Dynamism of Western Culture* (Cambridge, Mass.: MIT Press, 1968); Lynn White, Jr., "Science and the Sense of Self: The Medieval Background of a Modern Confrontation," *Daedalus* 107 (Spring 1978): 47–59.

7. White, *Machina Ex Deo*, p. 89.

8. Ibid., pp. 88–89.

9. Fernand Braudel, *The Structures of Everyday Life: The Limits of the Possible*, vol. 1 of *Civilization and Capitalism, 15th–18th Century*, trans. Sîan Reynolds (Berkeley: University of California Press, 1992), pp. 375–77, 399.

10. Sir Francis Bacon, *Novum Organum*, in *Francis Bacon, Great Books of the Western World*, aphorism 129, p. 135 (Chicago: Encyclopaedia Britannica, 1952). For analyses of the religious context for Bacon's work, see Robert N. Proctor, *Value-Free Science? Purity and Power in Modern Knowledge* (Cambridge, Mass.: Harvard University Press, 1991), chapter 2; Max Oelschlaeger, *The Idea of Wilderness* (New Haven, Conn.: Yale University Press, 1991), pp. 80–85; Marie Boas Hall, *The Scientific Renaissance 1450–1630* (New York: Dover Publications, 1994), pp. 249–51.

11. "[Nature] acts in [animals] according to the disposition of their organs, just as a clock which is only composed of wheels and weights is able to tell the hours and measure the time more correctly than we can do with all our wisdom." René Descartes, *Discourse on the Method of Rightly Conducting the Reason and Seeking for Truth in the Sciences*, in *Descartes/Spinoza, Great Books of the Western World* (Chicago: Encyclopaedia Britannica, 1952), p. 60.

12. Oelschlaeger, *The Idea of Wilderness*, p. 89.

13. Vannevar Bush, *Science, the Endless Frontier* (Washington, D.C.: Office of Scientific Research and Development, 1945; reprint, Washington, D.C.: National Science Foundation, 1960), p. 11.

14. For a discussion of historical views of the relation between values and science, see Proctor, *Value-Free Science*.

15. Leon M. Lederman, "The Advancement of Science," *Science* 256 (May 22, 1992): 1120.

16. Bacon, *Novum Organum*, aphorism 19, p. 108.

17. Michael Pravica, letter to *New York Times*, November 14, 1993, section 3, p. 11.

18. Silvan S. Schweber, "Physics, Community and the Crisis in Physical Theory," *Physics Today* (November 1993): 35, 39.

19. For a general discussion of emergence, and complexity science in general, see Mitchell M. Waldrop, *Complexity: The Emerging Science at the Edge of Order and Chaos* (New York: Simon and Schuster, 1992).

20. Rachel Carson, *Silent Spring* (Boston: Houghton Mifflin, 1962).

21. David Baltimore, "Limiting Science: A Biologist's Perspective," *Daedalus* 107 (Spring 1978): 42.

22. Václav Havel, "The New Measure of Man," *New York Times*, July 8, 1994, A27.

23. Daniel Kleppner, "Thoughts on Being Bad," *Physics Today* (August 1993): 9–10; Gerald Holton, "The Value of Science at the 'End of the Modern Era,'" speech delivered at February 1993 annual meeting of Sigma Xi, San Francisco, Calif., pp. 25–30.

24. Paul C. Stern, Oran R. Young, and Daniel Druckman, eds., *Global Environmental Change: Understanding the Human Dimensions* (Washington, D.C.: National Academy Press, 1992), p. 95; emphasis in original.

25. For example, see Marcia Barinaga, "A Bold New Program at Berkeley Runs into Trouble," *Science* 263 (March 11, 1994): 1367–68.

26. "How . . . can we adequately understand issues related to the environment unless we can draw upon the insights gained from studies in science and engineering, in economics and politics, in law and public administration, and in ethics and values (to name just a few of the relevant fields)? [We] will need to design our programs in a more integrated way [with] a closer link between research and practice, between classroom study and fieldwork, between academic analysis and the shaping of public policy." Neil L. Rudenstine, *The President's Report, 1991–1993* (Cambridge, Mass.: Harvard University, 1993), p. 7.

27. Steven Weinberg, "The Answer to (Almost) Everything," *New York Times,* March 8, 1993, A17.

28. Philip Anderson, quoted in Schweber, "Physics, Community, and the Crisis in Physical Theory," p. 36.

29. Jean Dausset and Howard Cann, "Our Genetic Patrimony," *Science* 264 (September 9, 1994): 1991.

30. Richard Strohman, "Epigenesis: The Missing Beat in Biotechnology," *Bio/Technology* 12 (February 1994): 156.

31. James E. Lovelock, "Gaia As Seen through the Atmosphere," *Atmospheric Environment* 6 (1972): 579–80; James E. Lovelock and Lynn Margulis, "Atmospheric Homeostasis by and for the Biosphere: The Gaia Hypothesis," *Tellus* 26 (1974): 2–10.

32. James E. Lovelock, "Geophysiology—The Science of Gaia," in *Scientists on Gaia,* ed. Stephen H. Schneider and Penelope J. Boston (Cambridge, Mass.: MIT Press, 1991), p. 4; emphasis added.

33. Lynn Margulis and Oona West, "Gaia and the Colonization of Mars," *GSA Today* 3 (November 1993): 277.

34. Charles Mann, "Lynn Margulis: Science's Unruly Earth Mother," *Science* 252 (April 19, 1991): 378–81.

35. George C. Williams, "Gaia, Nature Worship, and Biocentric Fallacies," *Quarterly Review of Biology* 67 (December 1992): 479.

36. Quoted in Timothy Ferris, *Coming of Age in the Milky Way* (New York: William Morrow, 1988), pp. 76–77.

37. Hall, *Scientific Renaissance,* chapter 10.

38. Paul Ehrlich, "Coevolution and Its Applicability to the Gaia Hypothesis," in Schneider and Boston, *Scientists on Gaia,* p. 21.

39. Descartes, *Discourse on Method,* part 4.

40. Lederman, "Advancement of Science," p. 1120.

41. Committee on Earth and Environmental Sciences, *Our Changing Planet: The FY 1994 U.S. Global Change Research Program, A Supplement to the U.S. President's Fiscal Year 1994 Budget* (Washington, D.C.: Committee on Earth and Environmental Sciences), p. 52.

42. Andy Johnson, "Science and Ethics," letter to the *New Yorker,* August 1, 1994, 7.

43. Schweber, "Physics, Community, and the Crisis in Physical Theory," pp. 39–40, emphasis in original. Schweber further notes that, although science differs from other "human practices" in that nature "places strong constraints" on the ways in which science can progress, this distinction does not diminish the moral connection between the creative process and the use of creative products in society (p. 40).

Chapter Seven: Pas de Trois: Science, Technology, and the Marketplace

1. Robert M. Solow, "Technical Change and the Aggregate Production Function," *Review of Economics and Statistics* (August 1957): 214–31.

2. This is not to say that everyone believes it. For a dissenting view see David F. Noble, "Automation Madness, or the Unautomatic History of Automation," in *Science, Technology, and Social Progress,* ed. Steven L. Goldman (Bethlehem, Pa.: Lehigh University Press, 1989), pp. 65–91.

3. Office of Management and Budget, *The Budget of the United States Government, Fiscal Year 1993* (Washington, D.C.: Executive Office of the President, 1992), part 1, p. 87.

4. William J. Clinton and Albert Gore, Jr., *Technology for America's Economic Growth: A New Direction to Build Strength* (Washington, D.C.: U.S. Government Printing Office, February 22, 1993), p. 7.

5. For a discussion of different ways to view standard of living, see essays by Amartya Sen and others in *The Standard of Living: The Tanner Lectures, Clare Hall, Cambridge, 1985,* ed. Geoffrey Hawthorn (New York: Cambridge University Press, 1987).

6. William J. Clinton and Albert Gore, Jr., *Science in the National Interest* (Washington, D.C.: Executive Office of the President, August 1994), introductory letter.

7. Rachel Nowak, "Genetic Testing Set for Takeoff," *Science* 265 (July 22, 1994): 464–67; Sherman Elias and George J. Annas, "Generic Consent for Genetic Screening," *New England Journal of Medicine* 330 (June 2, 1994): 1611–13.

8. Richard Strohman, "Epigenesis: The Missing Beat in Biotechnology," *Bio/*

Technology 12 (February 1994): 156–64; Charles C. Mann, "The Prostate-Cancer Dilemma," *Atlantic Monthly* (November 1993): 102–18; "Is a Little Knowledge a Dangerous Thing?" *The Economist* (August 21, 1993): 67–68; Ruth Hubbard and Elijah Wald, *Exploding the Gene Myth: How Genetic Information Is Produced and Manipulated by Scientists, Physicians, Employers, Insurance Companies, Educators, and Law Enforcers* (Boston: Beacon Press, 1993); also see essays on medical promise and social implications of genetic testing in *The Code of Codes: Scientific and Social Issues in the Human Genome Project*, ed. Daniel J. Kevles and Leroy Hood (Cambridge, Mass.: Harvard University Press, 1992).

9. Nathan Rosenberg and L. E. Birdzell, Jr., *How the West Grew Rich: The Economic Transformation of the Industrial World* (New York: Basic Books, 1986), p. 264.

10. For example, see Herbert A. Simon, *The Sciences of the Artificial* (Cambridge, Mass.: MIT Press, 1981), pp. 183–84.

11. The anthropologist Marvin Harris suggests that the earliest big-game hunters enjoyed "relatively high standards of comfort and security" and "an enviable standard of living," working only a few hours a day and enjoying the privilege of life at the top of the food chain: Marvin Harris, *Cannibals and Kings: The Origins of Cultures* (New York: Vintage Books, 1991), pp. 11–12.

12. Michael Polanyi, "The Republic of Science: Its Political and Economic Theory," *Minerva* 1 (Autumn 1962): 56, 55, 62.

13. Ernest L. Eliel, *Science and Serendipity: The Importance of Basic Research* (Washington, D.C.: American Chemical Society, March 1993), p. 3; also see National Science Foundation, *Benefits of Basic Research*, NSF 83–81 (Washington, D.C.: National Science Foundation, 1983).

14. James Gleick, "The Telephone Transformed—Into Almost Everything," *New York Times Magazine*, May 16, 1993, 26–29, 50–64.

15. The superhighway metaphor is perhaps more precise than its coiners intended. The interstate highway system, originally justified in terms of civil defense, has had an enormous but mixed effect on the evolution of America: solidifying the place of the automobile as the dominant transportation mode, opening up vast areas of the nation to development, compromising the economic viability of thousands of small towns, locking the nation into its dependence on oil, increasing the personal convenience of long-distance travel, destroying the commercial potential of long-distance passenger railroads.

For a variety of perspectives on the social and technological implications of the information superhighway, see "Seven Thinkers in Search of the Information Superhighway," *Technology Review* (August/September 1994): 42–52.

16. Andrew Pollack, "Now It's Japan's Turn to Play Catch-Up," *New York Times,* November 21, 1993, section 3, pp. 1, 6; also see "Second Time Around," *The Economist* (August 12, 1995): 51–52.

17. World Bank, *World Development Report 1994* (New York: Oxford University Press, 1994), p. 177.

18. Even the mainstream and risk-averse National Academy of Sciences has associated itself with this perspective; see Cheryl Simon Silver and Ruth S. De-Fries, *One Earth, One Future: Our Changing Global Environment* (Washington, D.C.: National Academy Press, 1990).

19. For example, see Julian L. Simon and Herman Kahn, eds., *The Resourceful Earth* (Cambridge, Mass.: Basil Blackwell, 1984).

20. For example, see the annual *State of the World* reports (New York: W. W. Norton) issued by the Worldwatch Institute; also see Robert D. Kaplan, "The Coming Anarchy," *Atlantic Monthly,* February 1994, 44–76.

21. These trends are also reflected in R&D priorities, of course. Before World War II, most government research was aimed specifically at meeting society's basic needs for commodities. In 1920, 40 percent of the federal civilian R&D budget was devoted to agriculture, and 25 percent went to exploration for and exploitation of natural resources; similar proportions held throughout the period between the world wars. Today, those values stand at about 4 and 6 percent, respectively (A. Hunter Dupree, *Science in the Federal Government: A History of Policies and Activities to 1940* [Cambridge, Mass.: Harvard University Press, 1957]).

22. World Bank, *World Development Report 1993* (New York: Oxford University Press, 1993), pp. 242–43; 256–57.

23. For a brief discussion of some basic democratic issues raised by science and technology in modern society, see Harold Lasswell, "Must Science Serve Political Power?" *American Psychologist* 25 (1971): 117–23. Some recent considerations of the problem include: Alvin M. Weinberg, "Technology and Democracy," *Minerva* 28 (1990): 81–90; David Guston, "The Essential Tension in Science and Democracy," *Social Epistemology* 7 (1993): 3–23; *Technology for the Common Good,* ed. Michael Shuman and Julia Sweig (Washington, D.C.: Institute for Policy Studies, 1993); *Technology in the Western Political Tradition,* ed. Arthur M. Melzer, Jerry Weinberger, and M. Richard Zinman (Ithaca, N.Y.: Cornell University Press, 1993).

24. Carnegie Commission on Science, Technology, and Government, *Partnerships for Global Development* (New York: Carnegie Commission on Science, Technology, and Government, 1992), p. 57.

25. United Nations Development Programme, *Human Development Report 1994*

(New York: Oxford University Press, 1994), pp. 129–218. There is, of course, no simple dividing line between "industrialized" and "developing" nations, and there is considerable heterogeneity within both groups (especially the latter). This report identifies thirty-three industrialized nations representing 22 percent of the world's population in 1990 and including much of the former Soviet Union.

26. Ibid.

27. World Bank, *World Development Report 1994*, pp. 224–25.

28. United Nations Development Programme, *Human Development Report 1994*. For an anecdotal account of the problem of concentration of technological capacity, see Mike Holderness, "Down and Out in the Global Village," *New Scientist* (May 8, 1993): 36–40.

29. Eugene B. Skolnikoff, *The Elusive Transformation: Science, Technology, and the Evolution of International Politics* (Princeton, N.J.: Princeton University Press, 1993), chapter 4; Enrique Martín del Campo, "Technology and the World Economy: The Case of the American Hemisphere," in *Globalization of Technology*, ed. Janet H. Muroyama and Guyford H. Stever (Washington, D.C.: National Academy Press, 1988), pp. 141–58; Francisco R. Sagasti, "Underdevelopment, Science and Technology: The Point of View of the Underdeveloped Countries," in *Views of Science, Technology and Development*, ed. Eugene Rabinowitch and Victor Rabinowitch (Oxford: Pergamon Press, 1975), pp. 41–53.

30. Dan L. Burk, Kenneth Barovsky, and Gladys Monroy, "Biodiversity and Biotechnology," *Science* 260 (June 25, 1993): 1900–1901.

31. Carnegie Commission on Science, Technology, and Government, *Science and Technology in U.S. International Affairs* (New York: Carnegie Commission on Science, Technology, and Government, 1992), p. 22.

32. United Nations Development Programme, *Human Development Report 1992* (New York: Oxford University Press, 1992), p. 40.

33. For example, see Nathan Rosenberg, *Exploring the Black Box: Technology, Economics, and History* (Cambridge: Cambridge University Press, 1994): 121–38.

34. For example, Robert Gilpin, *The Political Economy of International Relations* (Princeton, N.J.: Princeton University Press, 1987), pp. 303–5; John F. Devlin and Nonita T. Yap, "Sustainable Development and the NICs: Cautionary Tales for the South in the New World (Dis)Order," *Third World Quarterly* 15 (1994): 49–62.

35. United Nations Development Programme, *Human Development Report 1994*, pp. 142–43. This report adjusts per capita gross domestic product to reflect personal purchasing power, by compensating for "such factors as the degree

of openness of an economy . . . and possible overvaluation of exchange rates" (United Nations Development Programme, *Human Development Report 1993*, p. 109). "Real per capita GDP" includes a further adjustment for inequitable distribution of wealth within a nation.

36. Fernand Braudel, *The Wheels of Commerce*, vol. 2 of *Civilization and Capitalism, 15th–18th Century*, trans. Sian Reynolds (Berkeley: University of California Press, 1993), 507.

37. For a summary of the resource challenges facing the developing world, see World Commission on Environment and Development, *Our Common Future* (New York: Oxford University Press, 1987).

38. Thomas Malone, "Perspectives on Technology Transfer," the Harrelson Lecture, North Carolina State University, Raleigh, N.C., March 15, 1993, p. 3 (emphasis in original).

39. When useful products do emerge—such as satellite imagery—the costs of acquiring and analyzing the resulting data are often prohibitive.

40. United Nations Development Programme, *Human Development Report 1992*, p. 40.

41. For example, Skolnikoff, *Elusive Transformation*, chapter 4; Andrew Barnett, "Knowledge Transfer and Developing Countries: The Tasks for Science and Technology in the Global Perspective 2010, *Science and Public Policy* 21 (February 1994): 2–12; "While the Rich World Talks," *The Economist* (July 10, 1993): 11–12.

42. World Resources Institute, *World Resources 1994–95* (New York: Oxford University Press, 1994), chapter 4.

43. Charles Tilly, *Coercion, Capital, and European States, AD 990–1992* (Cambridge, Mass.: Basil Blackwell, 1992), p. 220.

44. For example, Ralf D. Hotchkiss, "Ground Swell on Wheels," *The Sciences* (July/August 1993): 14–18; Elizabeth Brubaker, "India's Greatest Planned Environmental Disaster," *Ecoforum* 14 (1989): 6–7.

45. Carnegie Commission on Science, Technology, and Government, *Partnerships for Global Development*, p. 95.

46. World Bank, *World Development Report 1993*, pp. 25, 152–53. "Global burden of disease" is a quantitative measure of public health encompassing "losses from premature death" and "loss of healthy life resulting from disability."

47. U.S. Congress, Office of Technology Assessment, *Status of Biomedical Research and Related Technology for Tropical Diseases*, OTA-H-258 (Washington, D.C.: U.S. Government Printing Office, September 1985); Tore Godal, "Fighting the Parasites of Poverty: Public Research, Private Industry, and Tropical Diseases," *Science* 264 (June 24, 1994): 1864–66.

48. Jon Cohen, "Bumps on the Vaccine Road," *Science* 265 (September 2, 1994): 1371–73.

Chapter Eight: Science as a Surrogate for Social Action

1. Daniel E. Koshland, Jr., "Science Advice to the President," *Science* 242 (December 16, 1988): 1489.
2. FASEB Public Affairs Executive Committee, "FASEB Response to Proposals for Reappraising Federal Support of Basic Biomedical Research," Federation of American Societies for Experimental Biology, February 1993, p. 3.
3. Office of Management and Budget, *The Budget of the United States Government, Fiscal Year 1992* (Washington, D.C.: Executive Office of the President, 1991), historical tables, pp. 37–48.
4. United Nations Development Programme, *Human Development Report 1994* (New York: Oxford University Press, 1994), pp. 186, 191; World Bank, *World Development Report 1993* (New York: Oxford University Press, 1993), pp. 292–93. Figures adjusted for purchasing power of local currency. Without the purchasing power adjustment—that is, in terms of absolute dollar expenditures—the discrepancies are even greater. In terms of relative expenditures, in 1991 the U.S. devoted 13.3 percent of its gross domestic product to health care; values for other industrialized nations include Sweden—8.8 percent; Norway—8 percent; Japan—6.8 percent; Portugal—5.4 percent; Greece—4.8 percent.
5. George H. Hitchings, "Health Care and Life Expectancy," letter to *Science* 262 (December 10, 1993): 1634. The author is with Burroughs Wellcome Co.
6. Michael E. DeBakey, "Medical Centers of Excellence and Health Reform," *Science* 262 (October 22, 1993): 523.
7. Leonard A. Sagan, *The Health of Nations* (New York: Basic Books, 1987), chapter 1; Thomas McKeown, *The Origins of Human Disease* (Cambridge, Mass.: Basil Blackwell, 1988), chapter 3.
8. McKeown, *Origins of Human Disease*, pp. 72–74; World Bank, *World Development Report, 1993*, pp. 292–93.
9. McKeown, *Origins of Human Disease*, p. 77.
10. John B. McKinlay and Sonja M. McKinlay, "The Questionable Contribution of Medical Measures to the Decline of Mortality in the United States in the Twentieth Century," *Milbank Memorial Fund Quarterly* 55 (Summer 1977): 405–27; Sagan, *Health of Nations*, chapter 4; McKeown, *Origins of Human Disease*, pp. 78–81.
11. McKinlay and McKinlay, "Questionable Contribution," especially figure 2. Available data suggest that from 1850 to 1900 the annual increase in life

expectancy in the industrialized world averaged about 0.18 years; from 1900 to 1950 this rose to almost 0.4 years, and from 1950 to 1992, fell to about 0.16 years (Sagan, *Health of Nations,* p. 16; United Nations Development Programme, *Human Development Report 1994,* p. 187).

12. Victor R. Fuchs, "Economics, Health, and Post-Industrial Society," *Milbank Memorial Fund Quarterly* 57 (Spring 1979): 155.

13. Sagan, *Health of Nations,* chapters 1, 4; McKinlay and McKinlay, "Questionable Contribution."

14. Eliot Marshall, "Experts Clash over Cancer Data," *Science* 250 (November 16, 1990): 900–902; Tim Beardsley, "A War Not Won," *Scientific American* (January 1994): 130–38.

15. McKeown, *Origins of Human Disease;* Sagan, *Health of Nations;* Fuchs, "Economics, Health, and Post-Industrial Society."

16. Kenneth D. Kochanek, Jeffrey D. Maurer, and Harry M. Rosenberg, "Why Did Black Life Expectancy Decline from 1984 through 1989 in the United States?" *American Journal of Public Health* 84 (June 1994): 938–44; J. S. Feinstein, "The Relationship between Socioeconomic Status and Health: A Review of the Literature," *Milbank Memorial Fund Quarterly* 71 (Spring 1993): 279–322.

17. Beardsley, "A War Not Won," pp. 137–38; "Is a Little Knowledge a Dangerous Thing?" *The Economist* (August 21, 1993): 67–68.

18. World Bank, *World Development Report 1993,* pp. 292–93; United Nations Development Programme, *Human Development Report 1993,* p. 26; "Infant Mortality Down; Race Disparity Widens," *Washington Post,* March 12, 1993, A6; Mark Schlesinger and Karl Kronebusch, "The Baby and the Bathwater: Failing Public Policies and Prenatal Care for Low-Income Women," John F. Kennedy School of Government, Working Paper Series H-90–4 (Cambridge, Mass.: Malcolm Wiener Center for Social Policy, Harvard University, 1990).

19. DeBakey, "Medical Centers of Excellence," p. 523.

20. Daniel E. Koshland, Jr., "Sequences and Consequences of the Human Genome," *Science* 246 (October 13, 1989): 189.

21. DeBakey, "Medical Centers of Excellence," p. 523.

22. World Bank, *World Development Report 1993,* chapters 3 and 4.

23. Richard Jerome, "Whither Doctors? Whence New Drugs?" *The Sciences* (May/June 1994): 23.

24. A 1992 poll indicated that 79 percent of Americans believed that the United States was devoting too little money to "improving health care." (National Science Board, *Science and Engineering Indicators—1993,* NSB 93–1 [Washington, D.C.: U.S. Government Printing Office, 1993], p. 489.)

25. Fuchs, "Economics, Health, and Post-Industrial Society," p. 165.

26. Andrew Lawler, "New GOP Chairs Size up Science," *Science* 266 (December 16, 1994): 1796–97; "House Starts Process of Cutting R&D Spending," *Science and Government Report* 25 (July 15, 1995): 1.

27. Jon Cohen, "Bumps on the Vaccine Road," *Science* 265 (September 2, 1994): 1371–73; Robert F. Allnutt, executive vice president of the Pharmaceutical Manufacturers Association, in "It's Health Care, Stupid!" *The Sciences* (January/February 1994): 20; Hitchings, "Health Care and Life Expectancy," p. 1634.

28. Elisabeth Rosenthal, "Drug Companies' Profits Finance More Promotion Than Research," *New York Times,* February 21, 1993, 1, 26.

29. Paul C. Stern, Oran R. Young, and Daniel Druckman, eds., *Global Environmental Change: Understanding the Human Dimensions* (Washington, D.C.: National Academy Press, 1992), chapter 4; Nathan Rosenberg, *Exploring the Black Box: Technology, Economics, and History* (New York: Cambridge University Press, 1994), chapter 9.

30. From 1972 to 1985, energy use increased by an average of only 0.3 percent a year, while energy intensity decreased an average of 2.4 percent each year; in the late 1980s, energy use increased by about 2 percent each year, while energy intensity decreased an average of only 0.8 percent per year. For example, see U.S. Congress, Office of Technology Assessment, *Energy Technology Choices: Shaping Our Future,* OTA-E-493 (Washington, D.C.: U.S. Government Printing Office, July 1991), chapter 1; U.S. Congress, Office of Technology Assessment, *Changing by Degrees: Steps to Reduce Greenhouse Gases,* OTA-O-482 (Washington, D.C.: U.S. Government Printing Office, February 1991), chapters 3 and 6.

31. Fred J. Sissine, "Energy Conservation: Technical Efficiency and Program Effectiveness," *Issue Brief,* IB85130 (Washington, D.C.: Congressional Research Service, April 13, 1993).

32. Linda R. Cohen and Roger G. Noll, "Synthetic Fuels from Coal," in *The Technology Pork Barrel,* ed. Linda R. Cohen and Roger G. Noll (Washington, D.C.: Brookings Institution, 1991), pp. 259–319.

33. Sissine, "Energy Conservation."

34. U.S. Department of Energy, *National Energy Strategy: Powerful Ideas for America,* DOE/S-0082P (Washington, D.C.: U.S. Government Printing Office, February 1991), p. C21.

35. Ibid., p. C22.

36. *Energy Policy Act of 1992,* Public Law 102–486, 102d Cong., 2d sess. (October 24, 1992).

37. Rosenberg, *Exploring the Black Box,* pp. 183–84 (emphasis in original).

38. For example, Lester C. Thurow, *The Zero-Sum Society* (New York: Penguin Books, 1980), chapter 2.

39. Jesse H. Ausubel, "Directions for Environmental Technologies," *Technology in Society* 16 (1994): 139–54.

40. United Nations Development Programme, *Human Development Report 1994,* p. 179.

41. Ibid.

42. U.S. Congress, Office of Technology Assessment, *Energy in Developing Countries,* OTA-E-486 (Washington, D.C.: U.S. Government Printing Office, 1991), p. 78; also see "Energy Survey," *The Economist,* June 18, 1994 (18 pages).

43. Koshland, "Science Advice."

44. Amory B. Lovins and Hunter L. Lovins, "Reinventing the Wheels," *Atlantic Monthly,* January 1995, 85.

45. For mainstream views, see Carnegie Commission on Science, Technology, and Government, *Risk and the Environment: Improving Regulatory Decision Making* (New York: Carnegie Commission on Science, Technology, and Government, June 1993), and Committee on Risk Assessment Methodology, *Issues in Risk Assessment* (Washington, D.C.: National Academy Press, 1993).

46. Langdon Winner, *The Whale and the Reactor: Searching for Limits in an Age of High Technology* (Chicago: University of Chicago Press, 1986), chapter 8.

47. Paul E. Gray, "The Paradox of Technological Development," in *Technology and the Environment,* ed. Jesse H. Ausubel and Hedy E. Sladovich (Washington, D.C.: National Academy Press, 1989), p. 200.

48. Nor do such assessments easily achieve scientific consensus. Risk assessment and cost-benefit analysis are subject to the same failures of authoritativeness discussed in Chapter 5. For example, see Philip H. Abelson, "Risk Assessments of Low-Level Exposures," *Science* 265 (September 9, 1994): 1507, and response letters in *Science* 266 (November 18, 1994): 1141–45.

49. Bernadette Modell, "a leading medical geneticist from University College Hospital in London," quoted in Gail Vines, "The Hidden Cost of Sex Selection," *New Scientist* (May 1, 1993): 12–13.

50. For example, William K. Stevens, "What Really Threatens the Environment," *New York Times,* January 29, 1991, C4.

51. "Survey, The Future of Medicine," *The Economist* (March 19, 1994): 13.

52. Ibid., p. 15.

53. Ibid., p. 18.

54. Virginia Morell, "Evidence Found for a Possible Aggression Gene," *Science* 260 (June 18, 1993): 1722–23.

55. Daniel E. Koshland, Jr., "The Dimensions of the Brain," *Science* 258 (October 9, 1992): 199.

56. Daniel E. Koshland, Jr., "Frontiers in Neuroscience," *Science* 262 (October 29, 1993): 635.

57. Daniel E. Koshland, Jr., "The Rational Approach to the Irrational," *Science* 250 (October 12, 1990): 189.

58. For example, see Monika Renneberg and Mark Walker, eds., *Science, Technology, and National Socialism* (New York: Cambridge University Press, 1993); Loren Graham, *Science in Russia and the Soviet Union* (New York: Cambridge University Press, 1993).

59. Albert J. Reiss, Jr., and Jeffrey A. Roth, eds. *Understanding and Preventing Violence* (Washington, D.C.: National Academy Press, 1993); Jane Ellen Stevens, "The Biology of Violence," *BioScience* 44 (May 1994): 291–94.

60. Fred Hirsch, *Social Limits to Growth* (Cambridge, Mass.: Harvard University Press, 1976).

Chapter Nine: Toward a New Mythology

1. For thoughtful perspectives on this turmoil, see Wil Lepkowski, "Science-Technology Policy Seems Set for New Directions in Clinton Era," *Chemical and Engineering News* (December 7, 1992): 7–14; Roland W. Schmitt, "Public Support of Science: Searching for Harmony," *Physics Today* (January 1994): 29–33; Radford Byerly, Jr., and Roger A. Pielke, Jr., "The Changing Ecology of United States Science," *Science* 269 (September 15, 1995): 1531–32.

2. For example, see George E. Brown, Jr., "Academic Earmarks: An Interim Report of the Chairman of the Committee on Science, Space, and Technology," August 9, 1993 (unpublished report).

3. For example, Committee on Science, Engineering, and Public Policy, *Science, Technology, and the Federal Government: National Goals for a New Era* (Washington, D.C.: National Academy Press, 1993); William J. Clinton and Albert Gore, Jr., *Technology for America's Economic Growth: A New Direction to Build Strength* (Washington, D.C.: U.S. Government Printing Office, February 22, 1993); William J. Clinton and Albert Gore, Jr., *Science in the National Interest* (Washington, D.C.: Executive Office of the President, August 1994); and National Science Board, *A Foundation for the 21st Century: A Progressive Framework for the National Science Foundation, A Report of the National Science Board Commission on the Future of the National Science Foundation* (Washington, D.C.: National Science Foundation, November 20, 1992).

The most comprehensive approach has been the work of the Carnegie Commission on Science, Technology, and Government, which, between 1988 and 1994, issued nineteen reports on various aspects of science and technology policy. Most of these reports could be characterized as recipes for reorganization, and few of them offer significant challenges to the status quo. Two partial exceptions are the reports *Partnerships for Global Development: The Clearing Horizon* (New York: Carnegie Commission on Science, Technology, and Government, December 1992), which seeks to articulate a global vision for science and technology policy, and *Enabling the Future: Linking Science and Technology to Societal Goals* (New York: Carnegie Commission on Science, Technology, and Government, September 1992), which recognizes the need for meaningful dialogue between the public and the R&D community.

4. For example, Clinton and Gore, *Science in the National Interest;* this report begins with a quotation from Vannevar Bush and shortly thereafter reiterates, "Science does indeed provide an endless frontier. Advancing that frontier and exploring the cosmos that we live in helps to feed our sense of adventure and our passion for discovery. Science is also an endless resource: in advancing the frontier, our knowledge of the physical and living world constantly expands" (p. 1).

5. National Science Board, *Science and Engineering Indicators—1993*, NSB 93-1 (Washington, D.C.: U.S. Government Printing Office, 1993), pp. 284–86.

6. A critique of this isolation is presented in Bruce Bimber and David Guston, "Politics by the Same Means," in *Handbook of Science and Technology Studies*, ed. Sheila Jasanoff, Gerald E. Markle, James C. Petersen, and Trevor Pinch (London: Sage Publications, 1994), pp. 554–71.

7. This is by no means a new observation; for example, see Emmanuel G. Mesthene, "The Impacts of Science on Public Policy," *Public Administration Review* (June 1967): 97–104.

8. Paul C. Stern, "A Second Environmental Science," *Science* 260 (June 25, 1993): 1897. Also see Paul C. Stern and Elliot Aronson, eds., *Energy Use: The Human Dimension* (New York: W. H. Freeman, 1984).

9. For example, see discussion of DNA fingerprinting in Chapter 5.

10. For example, see Committee on Risk Perception and Communication, *Improving Risk Communication* (Washington, D.C.: National Academy Press, 1989).

11. For example, see Paul Slovic, "Perceived Risk, Trust, and Democracy," *Risk Analysis* 13 (1993): 675–82.

12. John R. Steelman, *Science and Public Policy* (Washington, D.C.: U.S. Government Printing Office, 1947), p. viii.

13. Robert Chambers, "Beyond the Green Revolution: A Selective Essay," in *Understanding Green Revolutions*, ed. Tim P. Bayliss-Smith and Sudhir Wanmali (Cambridge: Cambridge University Press, 1984), p. 374.

14. Paul C. Stern, Oran R. Young, and Daniel Druckman, eds., *Global Environmental Change: Understanding the Human Dimensions* (Washington, D.C.: National Academy Press, 1992), pp. 168–70.

15. The potential market for the cystic fibrosis test alone has been estimated as somewhere between $200 million and a billion dollars a year: Gina Kolata, "Rush Is on to Capitalize on Testing for Gene Causing Cystic Fibrosis," *New York Times*, February 6, 1990, C3. Recent identification of genetic sites for nonpolyposis colon cancer and breast and ovarian cancer promise a "huge potential market" for diagnostic testing: Rachel Nowak, "Genetic Testing Set for Takeoff," *Science* 265 (July 22, 1994): 464.

16. Walter Gilbert, "A Vision of the Grail," in *The Code of Codes: Scientific and Social Issues in the Human Genome Project*, ed. Daniel J. Kevles and Leroy Hood (Cambridge, Mass.: Harvard University Press, 1992), p. 84.

17. But see *The 21st Century Project: Setting a Course for Science and Technology Policy* (Palo Alto, Calif.: Computer Professionals for Social Responsibility, 1993).

18. Perhaps the best model for such an institute was the Office of Technology Assessment (OTA), but, as an arm of Congress, OTA was often conspicuously risk averse out of political necessity. These efforts to maintain a low profile did not prevent—and may even have hastened—OTA's demise at the hands of congressional budget cutters in 1995. Of course numerous nongovernmental policy institutes with expertise in research and development still exist, either as parts of larger, nonprofit "think tanks" such as the Brookings Institution, as parts of private companies such as the Rand Corporation, or as parts of academic institutions such as the Center for Science and International Affairs at Harvard's Kennedy School of Government. Such programs differ from what is being recommended here in that they do not occupy an explicitly intermediary and interactive position between policy makers and the R&D system, and they are usually not mandated to perform ongoing assessment and evaluation of particular R&D initiatives in the context of specific societal goals. On the other hand, the National Academy of Sciences, which sometimes does engage in assessment and evaluation of federal R&D programs, is in essence a surrogate for the R&D community itself and cannot be portrayed as an independent entity. Furthermore, to the extent that any of these organizations must constantly compete for federal funds for their continued survival (as in the case of the National Academy), they cannot be viewed as autonomous.

19. For example, see Edward S. Rubin, Lester B. Lave, and M. Granger Morgan, "Keeping Climate Research Relevant," *Issues in Science and Technology* (Winter

1991–92): 47–55; Ronald D. Brunner, "Policy and Global Change Research: A Modest Proposal," paper presented at the Fourteenth Annual Research Conference for the Association for Public Policy Analysis and Management, Washington, D.C., October 30, 1993; Roger A. Pielke, Jr., "Usable Information for Policy: An Appraisal of the U.S. Global Change Research Program," *Policy Sciences* 28 (1995): 39–77; also see Chapter 5.

20. *Joint Climate Project to Address Decisions Maker's Uncertainties, Report of Findings* (Washington, D.C.: Joint Climate Project, May 1992), p. 11.

21. Ibid., p. 13.

22. *Science for All Americans: A Project 2061 Report on Literacy Goals in Science, Mathematics, and Technology* (Washington, D.C.: American Association for the Advancement of Science, 1989), p. 12.

23. Ibid.

24. Daniel Yankelovich, *Coming to Public Judgment: Making Democracy Work in a Complex World* (Syracuse, N.Y.: Syracuse University Press, 1991); John Doble and Amy Richardson, "You Don't Have to be a Rocket Scientist," *Technology Review* (January 1992): 51–54; Dorothy Nelkin, "Threats and Promises: Negotiating the Control of Research," *Daedalus* 107 (Spring 1978): 191–209.

25. Carnegie Commission on Science, Technology, and Government, *Enabling the Future.*

26. See Gary Chapman, "The National Forum on Science and Technology Goals: Building a Democratic, Post–Cold War Science and Technology Policy," *Communications of the ACM* 37 (January 1994): 31–37. The National Science Foundation has recently provided start-up funding for the National Forum on Science and Technology Goals, which is now housed at the Lyndon Baines Johnson School of Public Affairs at the University of Texas. (Gary Chapman, personal communication, 1994.)

27. Yankelovich, *Coming to Judgment.*

28. Langdon Winner, "Citizen Virtues in a Technological Order," *Inquiry* 35 (1992): 341–61; Richard E. Sclove, "Technological Politics as if Democracy Really Mattered: Choices Confronting Progressives," in *Technology for the Common Good,* ed. Michael Schuman and Julia Sweig (Washington, D.C.: Institute for Policy Studies, 1993), pp. 54–79; Bernard Dixon, "Debating Biotechnology," *Bio/Technology* 12 (August 1994): 746.

29. Eugene B. Skolnikoff, *The Elusive Transformation: Science, Technology, and the Evolution of International Politics* (Princeton, N.J.: Princeton University Press, 1993), p. 135.

30. Matthew S. Gamser, "Innovation, Technical Assistance, and Development: The Importance of Technology Users," *World Development* 16 (1988): 714.

31. Amilcar O. Herrera quoted in ibid., p. 717.

32. Stephen J. Lansing, *Priests and Programmers: Technologies of Power in the Engineered Landscape of Bali* (Princeton, N.J.: Princeton University Press, 1991).

33. For example, see David W. Brokensha, D. M. Warren, and Oswald Werner, eds., *Indigenous Systems of Knowledge and Development* (Lanham, Md.: University Press of America, 1980); for a popularized example, see Jane Ellen Stevens, "Science and Religion at Work," *BioScience* 44 (February 1994): 60–64.

34. S. Varadarajan, "a leading Indian industrialist," quoted in: Fred Pearce, "The Hidden Cost of Technology Transfer," *New Scientist* (May 9, 1992): 38.

35. But see R. E. Johannes, *Words of the Lagoon* (Berkeley: University of California Press, 1981).

36. Quoted in Timothy Ferris, *Coming of Age in the Milky Way* (New York: William Morrow, 1988), p. 79.

37. Lynn White, Jr., *Machina Ex Deo: Essays in the Dynamism of Western Culture* (Cambridge, Mass.: MIT Press, 1968), p. 103.

38. Thomas S. Kuhn, *The Structure of Scientific Revolutions* (Chicago: University of Chicago Press, 1970), p. 68.

39. Ivan Amato, "A High-Flying Fix for Ozone Loss," *Science* 264 (June 3, 1994): 1410–12; Richard A. Kerr, "Iron Fertilization: A Tonic, but No Cure for the Greenhouse," *Science* 263 (February 25, 1994): 1089–90; Debora MacKenzie, "Will Tomorrow's Children Starve?" *New Scientist* (September 3, 1994): 24–29; Charles C. Mann, "Behavioral Genetics in Transition," *Science* 264 (June 17, 1994): 1686–89.

40. To be precise, one out of nineteen committee members was a woman.

41. Committee on Science, Engineering, and Public Policy, *Science, Technology, and the Federal Government*, pp. 29, 45.

42. Clinton and Gore, *Science in the National Interest*, p. 7.

43. Walter Lippmann, *Public Opinion* (New York: Free Press, 1922), p. 11.

44. Kenneth E. Boulding, *The Image: Knowledge in Life and Society* (Ann Arbor: University of Michigan Press, 1956), p. 122.

45. Sir Francis Bacon, *Novum Organum*, in *Francis Bacon, Great Books of the Western World*, aphorism 92, p. 125 (Chicago: Encyclopaedia Britannica, 1952).

46. Boulding, *The Image*, p. 78.

47. World Commission on Environment and Development, *Our Common Future* (New York: Oxford University Press, 1987), p. 43.

48. For example, see David Pearce, "Sustainable Futures: Some Economic Issues," in *Changing the Global Environment*, ed. Daniel B. Botkin, Margriet F. Cas-

well, John E. Estes, and Angelo A. Orio (San Diego, Calif.: Academic Press, 1989), pp. 311–23; Tom H. Tietenberg, "Managing the Transition: The Potential Role for Economic Policies," in *Preserving the Global Environment,* ed. Jessica Tuchman Mathews (New York: W. W. Norton, 1991), pp. 187–226; Robert Repetto, "Accounting for Environmental Assets," *Scientific American* (June 1992): 94–100.

49. In theory, alternative economic institutions could impose criteria of sustainability on the market system. For example, see Herman E. Daly, *Steady-State Economics* (Washington, D.C.: Island Press, 1991). The political obstacles to establishing such institutions, however, are immeasurably greater than those that inhibit change in the R&D system.

INDEX